STRANGE & UNEXPLAINED HAPPENINGS

When Nature Breaks the Rules of Science

STRANGE & UNEXPLAINED HAPPENINGS

When Nature Breaks the Rules of Science

volume 1

Jerome Clark and Nancy Pear

AN IMPRINT OF GALE RESEARCH,
AN INTERNATIONAL THOMPSON PUBLISHING COMPANY.

Changing the Way the World Learns

NEW YORK • LONDON • BONN • BOSTON • DETROIT • MADRID
MELBOURNE • MEXICO CITY • PARIS • SINGAPORE • TOKYO
TORONTO • WASHINGTON • ALBANY NY • BELMONT CA • CINCINNATI OH

STRANGE AND UNEXPLAINED HAPPENINGS:
When Nature Breaks the Rules of Science

Jerome Clark and Nancy Pear, *Editors*

STAFF

Sonia Benson, *U•X•L Developmental Editor*
Kathleen L. Witman, *U•X•L Associate Developmental Editor*
Carol DeKane Nagel, *U•X•L Managing Editor*
Thomas L. Romig, *U•X•L Publisher*

Margaret A. Chamberlain, *Permissions Associate (Pictures)*
Shanna P. Heilveil, *Production Associate*
Evi Seoud, *Assistant Production Manager*
Mary Beth Trimper, *Production Director*

Mary Krzewinski, *Art Director*
Cynthia Baldwin, *Product Design Manager*
Terry Colon, *Illustrator*

∞™ This book is printed on acid-free paper that meets the minimum requirements of American National Standard for Information Sciences—Permanence Paper for Printed Library Materials, ANSI Z39.48-1984.

ISBN 0-8103-9780-3 (Set)
    0-8103-9781-1 (Volume 1)
    0-8103-9782-X (Volume 2)
    0-8103-9889-3 (Volume 3)

Printed in the United States of America

U•X•L is an imprint of Gale Research Inc., an International Thomson Publishing Company. ITP logo is a trademark under license.

10 9 8 7 6 5 4 3 2 1

CONTENTS

VOLUME 1

I
UFOs: The Twentieth-Century
Mystery 1

II
Ancient ETs and Their
Calling Cards 23

XIX
Other Strange Events 487

READER'S GUIDE

Scope

Strange and Unexplained Happenings: When Nature Breaks the Rules of Science is a reliable reference guide to *physical* phenomena, as opposed to psychic or supernatural phenomena. It picks up where the studies of the occult and parapsychology leave off, telling the rest of the story of the world's mysteries—those dealing with strange natural and quasi-natural phenomena, things that seem to be a part of our world but are usually disputed or ignored by conventional science. No one who reads the newspaper or watches television is unfamiliar with reports of UFOs, the Loch Ness monsters, or Bigfoot. In *Strange* the stories of these and many more anomalies (abnormal or peculiar happenings) are presented with enough detail to stimulate wonder and surprise in even the most skeptical reader.

Research into such anomalies is not confined to a small group of nontraditional scientists. Academic teams, popular writers and members of anomalists' societies have examined strange physical phenomena in works ranging from sober, scientifically based analysis to wild conjecture. *Strange and Unexplained Happenings* presents the ideas of most of the major players in the field of each particular phenomenon, whether scientific or sensational. The entries also provide clear explanations of the kind of theory and research that have been applied. The greatest value of these fascinating accounts may not be in the answers they provide, but in the important questions they provoke.

Features

The three volumes of *Strange and Unexplained Happenings* are presented in chapters arranged by subject. Thus a student can look up Bigfoot and then browse the entire Shaggy, Two-footed Creatures in North America chapter, where he or she will find many other types of hairy bipeds that may or may not be related to the Bigfoot sightings. Similarly, by looking up "Flying Humanoids," the reader will find a variety of examples of visitors from other worlds.

Since anomalies, like most things in life, do not always fit neatly into one category or another, the volumes are extensively cross-referenced. Within each entry the names of related phenomena that have their own entries elsewhere in the set appear in boldface for quick reference. A thorough cumulative subject index concludes each volume.

The language used by anomalists (those who study strange happenings) presents a colorful variety of scientific and pseudoscientific terms. Although most of the terms are defined within the text, the glossary of terms appearing in the frontmatter of each volume ensures easy accessibility for all readers. Boxed material within the entry also provides interesting facts and explanations of concepts and terminology.

Strange centers on phenomena for which evidence is usually insufficient, if not altogether lacking. Photographs and drawings of all kinds have appeared as "proof" that some strange thing exists or happened. The photographs are often blurry and the drawings are often crude but, as the only evidence available, they are an important part of the history of the phenomenon and thus have been included in these volumes whenever possible.

Brief biographies of some of the foremost anomalists, whose works are most frequently cited within these volumes, are set off from the text in boxes. "Reel Life" boxes feature some modern and time-honored movies that have made such mysteries as werewolves, sea monsters, and UFOs a part of our culture.

Key sources are provided at the end of each entry. A Further Investigation section at the conclusion of each volume provides an annotated listing of the major books, periodicals, and organizations the student may wish to consult in further research into a particular strange phenomenon.

Comments and Suggestions

We welcome your comments on this work as well as your suggestions for strange events to be featured in future editions of *Strange and Unexplained Happenings*. Please write: Editors, *Strange and Unexplained Happenings*, U·X·L, 835 Penobscot Bldg., Detroit, Michigan 48226-4094; call toll-free: 1-800-877-4253; or fax: 313-961-6348.

INTRODUCTION

Strange and Unexplained Happenings: When Nature Breaks the Rules of Science is a book about *anomalies*, human experiences that go against common sense and break the rules that science uses to describe our world. In the words of folklorist Bill Ellis, "Weird stuff happens." *Strange and Unexplained Happenings* takes a look at the "weird stuff" that abounds in the reports of ordinary people who have had extraordinary experiences. Accounts of flying saucers, reptile men, werewolves, and abominable snowmen grab our attention and send shivers down our spines. When similar strange accounts are repeated time and again by different witnesses in different times and places, they capture the attention of the scientific community as well.

The three hardest words for human beings to utter are *I don't know.* Because we like our mysteries quickly and neatly explained, in modern times we have come to ask scientists to find logical explanations for strange human experiences. Sometimes science can use its knowledge and tools to find the answers to puzzling incidents; at other times it offers explanations that don't seem to fit the anomalies and only add to the confusion about them. When experiences are especially unbelievable, scientists may simply decide that they never really happened and refuse to consider them altogether. Most of us believe that as science learns more, it will be able to explain more. Still, it is almost certain that science will never be able to account for all the "weird stuff" that human beings encounter.

When an anomaly is reported, it is natural not to believe it, to be skeptical. One usually wonders about the witness. Could the person be lying for some reason? Tricks and hoaxes do occur. There are people who go to great lengths to fool scientists and the public, who hope to find fame and fortune by false claims or simply to prove to themselves how clever they are. Photographs of extraordinary happenings are often fake; it is thought that nearly 95 percent of all UFO photos are false, and some of the best film footage of the Loch Ness monster is judged doubtful as well. As a matter of fact, most investigators of anomalies feel that a lot of photographs of an incident signals a fake, because: (1) most people don't walk around with cameras ready to snap strange sights; (2) people having weird or scary experiences are often in shock or terror, and taking pictures is the last thing on their minds; and (3) anomalies generally last for just a matter of seconds. Investigators also believe that the fuzzier the photo, the more likely it is to be real, because pictures taken by people with shaking hands rarely turn out clearly, while hoaxers know that poor photographs won't get the results they are looking for.

It is also natural to wonder about the mental health of a person who has witnessed an extraordinary happening. Common sense tells us that *all* weird accounts should be blamed on the poor memories, bad dreams, or wild imaginings of confused and unwell minds! Still, psychologists who have examined witnesses of anomalies find them, for the most part, to be the same as people who have had no odd experiences at all. Also, the sheer number of strange reports rattles our common sense a bit, as do cases of multiple witnesses, when large groups of people observe the same strange happenings together.

More interesting still are accounts that have been repeated for centuries; reports of lake monsters in the deep waters of Loch Ness, for example, began way back in A.D. 565! Interesting, too, are reports that are widespread. The Pacific Northwest region of North America has its Bigfoot sightings, western Mongolians tell stories about the Almas, and accounts of the yeti have been reported in the high reaches of the Himalayas. While the languages and cultures surrounding these legends may differ, it is clear that witnesses are describing a similar creature: a hairy, two-legged "apeman." When observers report sightings of sea serpents they may describe them as smooth and snakelike or maned like horses, with many humps or finned like fish. Even when details vary widely, it is difficult to ignore their basic sameness: all suggest the existence of large, as yet unknown, sea-going animals.

It is true that in many cases of strange happenings, people have been misled or mistaken. The Bermuda Triangle, an area of the Caribbean where ships and planes were reported to mysteriously disappear, for example, was considered a real threat for more than two decades. That is, until weather records and other documents were properly researched, proving that the location was as safe as any other body of water. In the same way, the strange cattle "mutilations" that worried farmers in Minnesota and Kansas in the 1970s—and stirred up all sorts of weird explanations—required the special skills of veterinary pathologists to find that the cause was a simple, but gruesome, infection. Sometimes strange accounts do seem to change and grow as they are reported over the decades and in print. Human beings do want the truth ... but they also like a good story!

Science, too, has made its mistakes over the years. When sailors gave accounts of large sea creatures with giant eyes and many tentacles they were told that they were seeing floating trees with large roots. We now know that their accounts described giant squids. Gorillas and meteors were also rejected by scientists not that long ago!

But then, of course, some anomalies are more believable than others. When an odd happening turns the way we think about the world upside down it is described as "high-strange"; less weird accounts are lower on the strangeness scale. It is not *completely* unthinkable that unknown creatures still exist in some remote regions of the globe as many cryptozoologists (people who study "hidden" animals) believe. Wildlife experts and marine biologists may, over time, find that creatures like Bigfoot and "Nessie" are real. In the same way, physicists and meteorologists may find the reasons for ball lightning, or for the mysterious ice chunks that fall from the sky. The discovery of intelligent beings from outer space, on the other hand, would really shake us up and force us to rethink our lives and our place in the universe. As high on the strange scale as this idea is, though, there is enough hard evidence—like odd radar trackings and soil samples from UFO landing sights—to make it worth considering.

Strange accounts, no matter how farfetched, deserve some careful thought. Although most readers set their own limits as to how high on the strange scale they can go, the kinds of questions raised by anomalies are worth pursuing, even if the event or object is beyond one's own limits of belief. True understanding of anomalies takes time, effort, and an open—but not a gullible—mind. *Strange and Unexplained Happenings* doesn't deal with belief or disbelief. It only shows that human experiences come in more shapes and sizes than we could ever imagine!

Glossary

A

anomaly: something that is abnormal and difficult to explain or classify by conventional systems. An *anomalist* is someone who studies or collects anomalies.

anthropoid: ape.

anthropology: the study of human beings in terms of their social relations with each other, their culture, where they live, where they come from, physical characteristics, and their relationship with the environment.

archaeology: the scientific study of prehistory by finding and examining the remains of past life, such as fossils, relics, artifacts, and monuments.

arkeology: a term used to describe the search for the remains of Noah's Ark at the site where it landed after the Great Flood, as chronicled in the Book of Genesis in the Bible.

astronomy: the study of things that are outside of the Earth's atmosphere.

Atlantis: a fabled island in the Atlantic inhabited by a highly advanced culture. According to Greek legend an earthquake caused the island to be swallowed up by the sea. Some still believe in the legendary Atlantis today, and societies have arisen in order to actively search for its remains.

atmospheric life forms: *See space animals.*

B

bioluminescent organisms: plants and animals that make their own light by changing chemical energy into light energy. Bioluminescent organisms are especially common in places where no light penetrates, like the depths of the ocean.

bipeds: animals that walk on two feet.

C

CE1: a UFO seen at less than 500 feet from the witness.

CE2: a UFO that physically affects its surroundings.

CE3: a being observed in connection with a UFO sighting.

cereology: the study of crop circles.

coelacanth: a large fish that, until 1938, had only been known through fossil records and was thought to have been extinct for some 60 million years. In 1938 a coelacanth was caught in the net of a South African fishing boat, giving rise to speculation that other species that had been officially declared extinct may live on.

contactee: a person who claims to have ongoing communications with one or more extraterrestrials. A *physical contactee* claims to have had actual physical contact with extraterrestrials and often will produce photographs or other material evidence of these meetings. A *psychic contactee* claims to have received messages from space in dreams or through automatic writing (writing performed without thinking, seemingly directed by an outside force).

corpse candles: also called death-candles; lights appearing in the form of a flame or a luminous mass, according to folk tradition, that foretell an impending death.

cover-up: an attempt made by an organization or group to conceal from the public the group's actions or information it has received or collected.

creationism: the belief, based on a word-for-word reading of the Bible's Book of Genesis, that God created all matter, all living things, and the world itself, all at the same time and from nothing.

creation myths: sacred stories that explain how the Earth and its beings were created.

Cro-Magnon race: a race that lived 35,000 years ago and is of the same species as modern human beings (*Homo sapiens*). Cro-Magnons stood straight and were six or more feet tall; their foreheads were high and their brains large. Skillfully made Cro-Magnon tools, jewelry, and cave wall paintings suggest that the Cro-Magnon race had an advanced culture.

cryptozoology: the study of lore concerning animals that science does not account for, including animals thought to be extinct, animals that have been seen only by local populations, or animals thought to exist only in certain areas that show up elsewhere. The objective of cryptozoology is generally to evaluate the possibility of these animals' existence.

D

debunk: to expose something as false or as a hoax.

dowsing: a folk method for finding underground water or minerals with a divining rod. The divining rod is usually a forked twig; the "diviner" holds the forked ends close to his or her body, and the stem supposedly points downward when he or she walks over the hidden water or desired mineral. Some believe that dowsing can be used as a method to predict when and where a crop circle will appear.

E

ethereans: fourth dimensional human beings; a theoretical group of beings like humans, only more advanced, who live in another (or fourth) dimension that coexists with our world. Just as the stars and planets of our universe have their etheric counterparts, ethereans are human beings in a different reality.

evolution: a process in which a group of plants or animals—such as a species—changes over a long period of time, so that descendants differ from their ancestors. Theoretically the changes result from *natural selection,* a process in which the strongest and the most adept at survival pass on their characteristics to the next generations. Characteristics that make group members less successful at surviving and breeding are slowly weeded out.

extinct: no longer in existence.

extraterrestrial: something that came into being or lives outside of the Earth's atmosphere, or something that happened outside of the Earth's atmosphere.

F

Fortean: an adjective used to describe outlandish, sometimes sarcastic, and generally antiscientific theories in the literature of the strange and unexplained. The word is derived from the pioneer of physical anomalies, Charles Fort, who frequently poked fun at the weak attempts science made to explain away strange events by offering wacky theories of his own.

G

geophysics: a branch of earth science dealing with physical processes and phenomena occurring within or on the Earth.

H

hallucination: an illusion of seeing, hearing, or in some way becoming aware of something that apparently does not exist in reality.

herpetology: the scientific study of reptiles and amphibians.

humanoid: having human characteristics; a being that resembles a human.

I

ichthyology: the study of fish.

inorganic: composed of matter other than plant or animal; relating to mineral matter as opposed to the substance of things that are or were alive.

L

Lemuria: a legendary lost continent in the Pacific Ocean somewhere between southern Africa and southern India. Unlike accounts of Atlantis, which date back to the writings of Plato in ancient Greece, theories about Lemuria arose in the nineteenth century in the doctrine of occultists such as Madame Helena Petrovna Blavatsky, cofounder of the Theosophical Society, Max Heindel, founder of the Rosicrucian Fellowship, Rudolf Steiner, founder of the Anthroposophical Society, and Theosophist W. Scott-Elliott.

lycanthropy: the transformation of a man or woman into a wolf or wolflike human.

M

meteor: one of the small pieces of matter in the solar system that can be seen only when it falls into the Earth's atmosphere, where friction may cause it to burn or glow. When this happens it is sometimes called a "falling" or "shooting" star.

meteorite: a meteor that survives the fall to Earth.

meteoroid: any piece of matter—ranging in mass from a speck of dust to thousands of tons—that travels through space; it is composed largely of stone or iron or a mixture of the two. When a meteoroid enters the Earth's atmosphere it becomes visible and is called a *meteor*.

meteorology: the science of weather and other atmospheric phenomena.

mollusks: the second largest group of invertebrate animals (those without a backbone). They are soft-bodied, and most have a distinct shell. Mollusks usually live in water and include scallops, clams, oysters, mussels, snails, squids, and octopuses.

mutology: the investigation of cattle mutilations.

N

Neanderthal race: a species that lived between 40,000 and 100,000 years ago. Neanderthal remains have been found in Europe, northern Africa, the Middle East, and Siberia. The classic Neanderthal man had a large thick skull with heavy brow ridges, a sloping forehead, and a chinless jaw. He was slightly over five feet tall and had a stocky body. The link between Neanderthal man and modern human beings is unclear. Many anthropologists believe that they evolved separately from an earlier common ancestor.

New Age: a late-twentieth-century social movement that draws from American Indian and Eastern traditions and espouses spirituality, holism, concern for the environment, and metaphysics.

O

occultism: belief in or study of supernatural powers.

OINTS (Other Intelligences): a term coined by biologist Ivan T. Sanderson that includes not only extraterrestrials, but also space animals, undersea civilizations, poltergeists, and extradimensional beings.

organic: composed of living plant or animal matter.

ornithology: the scientific study of birds.

P

paleontology: the scientific study of the past through fossils and ancient forms of life.

paracryptozoology: the study of animals whose existence, even to the most open-minded, seems impossible (*paracryptozoology* means "beyond cryptozoology").

paranormal: not scientifically explainable; supernatural.

paraphysical: a combination of the terms *paranormal* (outside the normal) and *physical*. This concept, used by some anomalists, encompasses both natural occurrences (like leaving tracks) and unnatural occurrences (like disappearing instantly).

plesiosaurs: a suborder of prehistoric reptiles that dominated the seas during the Cretaceous Period (136 to 65 million years ago). Their bodies were short, broad, and flat. They had short pointed tails. Their small heads were supported by long slender necks, ideal for darting into the water to catch fish. Plesiosaurs swam with a rowing movement, using their four powerful, diamond-shaped flippers like paddles. They were often quite large, measuring up to 40 feet in length.

porphyria: a rare genetic disease often linked with werewolf sightings. Porphyria sufferers are plagued by tissue destruction in the face and fingers, open sores, and extreme sensitivity to light. Their facial skin may take on a brownish cast, and they may also suffer from mental illness. The inability to tolerate light, plus shame stemming from physical deformities, may lead the afflicted to venture out only at night. Some in the medical community have suggested that sightings of werewolves have really been of individuals with porphyria.

primates: a member of the group of mammals that includes man, apes, monkeys, and prosimians, or lower primates. Primates have highly developed brains and hands with opposable thumbs that are very adept at holding and grasping things.

psychosocial hypothesis: a belief that UFOs and other anomalies are powerful hallucinations shaped by the witness's psyche and culture, and that strange sightings are actually insights into deep realms of the human imagination rather than evidence of visitors from other worlds.

Q

quadrupeds: animals that walk on four feet.

R

radar: a method of detecting distant objects and determining their position, speed, or other characteristics by analyzing radio waves reflected from them.

S

saucer nests: circular indentations that one could imagine to be left by hovering or grounded spacecraft; saucer nests, found in the 1960s and 1970s before the current crop circle mania began, have many features comparable to crop circles, but the connection between the two phenomena has not been established.

sauropods: huge plant-eating reptiles with long necks and tails, small heads, bulky bodies, and stumplike legs; *Diplodocus, Apatosaurus (Brontosaurus),* and *Brachiosaurus* were sauropods.

shape-shifter (or shape-changer): one who can change form at will; in medieval and later chronicles, shape-shifting was associated with witchcraft, and such shape-shifters as black dogs and werewolves were often considered to be either agents of the devil or Satan himself.

sonar: a method of tracking that uses reflected sound waves to detect and locate underwater objects.

space animals (atmospheric life forms): hypothetical life forms existing in the upper atmosphere. Several ufologists have suggested that UFOs are neither spacecraft nor cases of mistaken identity but *space animals.*

spontaneous generation: a once widespread belief that living things can spring from nonliving material; thus, when rain hits the ground it can give rise—out of the mud, slime, and dust—to all sorts of living matter.

supernatural: of or relating to an order of existence outside the natural, observable universe.

T

teleportation: the act of moving an object or person from one place to another by using the mind, without using physical means.

theosophy: teachings about God and the world based on mystical insight; in 1875 the Theosophy movement arose in the United States, following Eastern theories of evolution and reincarnation.

transient lunar phenomena (TLP): unusual, short-lasting appearances on the moon's surface typically observed by astronomers through telescopes, and more rarely by the naked eye.

U

UFO (unidentified flying object): a term first coined by a U.S. Air Force worker that came into common usage in the mid-1950s to describe the "flying saucers" or mysterious discs that were being observed in the air and were suspected by some to be the craft of extraterrestrial visitors.

UFO-abduction reports: the accounts of a significant group of witnesses who claim to have been kidnapped by aliens. Many witnesses described large-headed, gray-skinned humanoids who subjected them to medical examinations. Some witnesses experienced amnesia after their encounters and recalled them only through hypnosis.

ufology: the study of unidentified flying objects.

ultraterrestrials: beings from another reality.

W

water horse: a folkloric creature believed by many to be a dangerous shape-changer that can appear either as a shaggy man who leaps out of the dark onto the back of a lone traveler or as a young horse that, after tricking an unknowing soul onto its back, plunges to the bottom of the nearest lake, killing its rider.

waterspouts: funnel- or tube-shaped columns of rotating, cloud-filled wind, usually extending from a cloud down to the spray it tears up from the surface of an ocean or lake.

Z

zeuglodon: a primitive, snakelike whale thought to have become extinct long ago.

PICTURE CREDITS

STRANGE & UNEXPLAINED HAPPENINGS

When Nature Breaks the Rules of Science

UFOs: The Twentieth-Century Mystery

- **UNIDENTIFIED FLYING OBJECTS**
- **UNIDENTIFIED AIRSHIPS**

UFOs: The Twentieth-Century Mystery

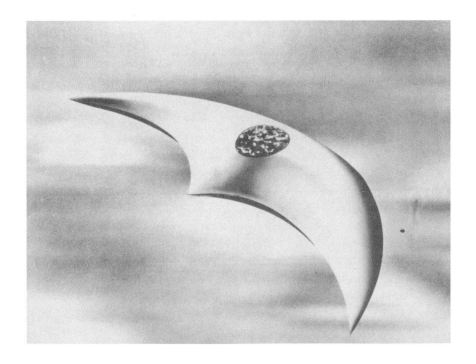

One of the nine "flying saucers" Kenneth Arnold sighted.

UNIDENTIFIED FLYING OBJECTS

One January evening in 1878, as he was hunting six miles south of Denison, Texas, John Martin saw a fast-moving object in the southern sky. When it passed overhead, he noted that it looked like a "large saucer." It would be a sight that many others would report in the years ahead, especially in the second half of the twentieth century.

"Flying saucers" were not really perceived or described as such until June 24, 1947, when a pilot named Kenneth Arnold spotted nine disc-shaped objects flying in formation at an estimated 1,200 mph over Mount Rainier, Washington. In a newspaper interview he compared their motion to that of stones skipping across water. Soon afterward, a newspaper in the area used the phrase "flying saucer" to describe what Arnold had seen—and the UFO age had begun. The more practical term "unidentified flying objects," or UFOs, first coined by a U.S. Air Force worker, would not come into common usage until the mid-1950s.

Yet, as demonstrated in the Martin account above, sightings of flying saucers and UFOs occurred well before 1947; they were probably just reported differently. From November 1896 until May 1897, for example, newspapers all across America were filled with stories about mysterious "airships"—cigar-shaped objects often said to flash bright searchlights and thought by some to be linked to visitors from Mars (also see entry: **Strange Clouds**). A few reports of what could be called UFOs can be found in scientific journals and newspapers of the earlier decades of the nineteenth century but are rare before that. For whatever reason, UFO sightings seem to be relatively recent phenomena.

Foo Fighters and Ghost Rockets

During World War II Allied pilots both in Europe and the Pacific observed many unidentified flying objects; they called them "foo fighters" and assumed they were enemy devices. When mysterious "ghost rockets" were sighted all over northern Europe during much of 1946, the Soviets were falsely blamed by observers who struggled to explain them.

Project Sign

The first U.S. Air Force effort to study UFO reports was conducted under the code name Project Sign and was established under the Air Materiel Command at Wright Field, Dayton, Ohio (later Wright-Patterson Air Force Base), on December 30, 1947. Routine sightings were usually handled at local air bases, but Sign investigated reported sightings considered important or unusual.

The first of these investigations was a January 7, 1948, case in which a Kentucky Air National Guard pilot, Captain Thomas F. Mantell, Jr., died in a plane crash while trying to check something that he described in one of his last radio messages as a "metallic object ... of tremendous size." The air force eventually identified the "object" as an air balloon connected with the navy's then-secret Skyhook Project.

A more puzzling report came later in 1948 from two Eastern Airlines pilots, Clarence S. Chiles and John B. Whitted. As their DC-3 flew over Alabama at 2:45 A.M. on July 24, Chiles and Whitted saw a wingless, torpedo-shaped object streak past them. It had two rows of square windows from which, Chiles reported, "a very bright light was glowing. Underneath the ship there was a blue glow of light." A flame extended 50 feet from the rear. Although the UFO was in view for no more than ten seconds, it was also seen by a passenger. Sign investigators also learned that an hour earlier, a ground-maintenance crewman at Robins Air Force Base in Georgia had seen an identical UFO. Weirder still, four days before that, a rocket-shaped object with two rows of windows had been seen over The Hague, Netherlands!

By the time of this sighting, Project Sign had split into several groups of investigators, each with a different view concerning UFOs. One group believed that the objects were spacecraft from other worlds, while another thought they were Soviet secret weapons, and still another felt they were common objects—the identity of which had somehow been confused. In the case of the Chiles-Whitted sighting, the first group of investigators won out: a top-secret report arguing that the UFO evidence pointed to otherworldly visitation was sent all the way to air force chief of staff General Hoyt S. Vandenberg. It was not the conclusion Vandenberg wanted to hear, and he ordered all copies of the report burned. The document remained secret until 1956, when a book by a retired air force UFO-project officer, Edward J. Ruppelt, reported the story behind it. Though other sources backed Ruppelt's account, the air force continued to deny that the report existed for many years.

Project Grudge

Vandenberg's rejection of Project Sign's conclusion sent a clear message to the investigators; those who believed in the possibility of extraterrestrial visitors either left the air force or were reassigned to other duties. On February 11, 1949, Project Grudge replaced Project Sign, and most UFO investigations then consisted simply of "debunking"—demonstrating that sightings and reports revealed nothing unusual, that they were really distortions or mistakes. By the end of the year the project's administrators had put most of its files into storage,

A top-secret report arguing that the UFO evidence pointed to otherworldly visitation was not what air force chief of staff General Hoyt S. Vandenberg wanted to hear. He ordered all copies of the report burned.

A UFO photographed by George J. Stock at Passaic, New Jersey, in 1952—one of a series of five photographs.

and by the following summer it had dwindled to a single investigator.

Project Blue Book

High-ranking air force officials called for a reorganization of Project Grudge, though, after its poor investigation of a series of radar/visual sightings of fast-moving UFOs over Fort Monmouth, New Jersey, in September 1951. Project Blue Book replaced it in March 1952, headed by Lieutenant Ruppelt, an intelligence officer assigned to the Air Technical Intelligence Center (ATIC) at Wright-Patterson Air Force Base. Ruppelt insisted that his investigators hold no prior judgments about whether or not UFOs were real. By the time he left the project two years later, Ruppelt was largely convinced that space visitors did exist. His memoir of his experiences, *The Report on Unidentified Flying Objects* (1956), is considered one of ufology's (the study of UFOs) most important books.

After Ruppelt left, however, the project returned to its former pattern of debunking—not investigating. Such was the case when a spectacular series of UFO radar and eyewitness observations occurred over Washington, D.C., in July 1952. Government intelligence officials became concerned that the Soviet Union could somehow use the sightings to cause mass panic in the United States, and they set up a panel of five scientists to secretly examine Blue Book's data and devise a security strategy.

The Robertson Panel

Over the next four days, the five scientists studied a few sighting reports and two UFO films before declaring further official study a "great waste of effort." The Robertson panel (named after its head, physicist and CIA employee H. P. Robertson) also called for a public "debunking" campaign that "would result in the reduction of public interest in 'flying saucers.'" In addition, it urged that UFO groups made up of ordinary citizens "be watched because of their potentially great

influence on mass thinking," stipulating that the "possible use of such groups for subversive purposes should be kept in mind."

Though the Robertson panel and its recommendations remained secret for years, they would have a huge effect on the course of UFO history. The air force began almost immediately to reduce Project Blue Book's funds and importance, and the program became devoted to downplaying sightings. Even air force chief scientific adviser J. Allen Hynek, who had attended the project's meetings, complained, "The Robertson panel ... made the subject of UFOs scientifically unrespectable, and for nearly 20 years not enough attention was paid to the subject to acquire the kind of data needed even to decide the nature of the UFO phenomenon."

Air Force Ignores Its Own Think Tank

A similar official cover-up took place following the publication of *Project Blue Book Special Report 14* in 1955. The report contained the findings of a three-year study by the Battelle Memorial Institute, a

Close-up of UFO photographed by George J. Stock, New Jersey, 1952.

Josef Allen Hynek

(1910-1986)

Josef Allen Hynek called himself the "innocent bystander who got shot"—a respected astronomer who became, quite by accident, the world's leading expert on UFOs. Educated at the University of Chicago and a popular writer on astronomy, he was assigned to the post of chief scientific adviser to Project Sign, the air force's UFO investigation project. He received the assignment simply because he was the astronomer nearest to Wright Field in Dayton, Ohio, where the project was based.

Beginning his work as air force consultant in 1948, Hynek at first doubted that UFOs were real. But as time went on, he could not help feeling puzzled by some of the reports he was receiving. By 1952 he admitted that amid all the false reports there might be a "residue that is worthy of scientific attention." Eventually, Hynek became a firm believer in the existence of UFOs. With Chicago businessman Sherman J. Larsen, he created the Center for UFO Studies (CUFOS) as a formal organization through which scientific research could be conducted.

think tank (a group of specialists in different areas of study who come together to try to solve complex problems) that the air force had asked to analyze UFO reports. Called Project Stork, the study concluded that UFOs were extraordinary occurrences, but that they did, in fact, exist. This was not what the air force wanted to hear! The report's data were drastically changed so that Secretary of the Air Force Donald A. Quarles could declare, "On the basis of this study we believe that no objects such as those popularly described as flying saucers have overflown the United States."

Because the air force always seemed to refuse to even consider the possibility of UFOs and because it often fabricated explanations, many feared that such debunking was a cover-up of real concerns. Perhaps, critics such as retired marine corps major Donald E. Keyhoe argued, the air force was well aware of the reality of space visitors but feared a worldwide panic if it admitted as much.

Eventually, the unreliability of Project Blue Book earned ridicule from the press and criticism from members of Congress. Finally, testifying before the House Armed Services Committee in April 1966, Hynek urged that a panel of physical and social scientists—this time unconnected with the government—"examine the UFO problem critically for the express purpose of determining whether a major problem really exists."

The Condon Committee: Study or Cover-up?

At this point the air force was only too eager to get the problem of UFO sightings off its hands. It asked the University of Colorado to conduct an independent scientific study. Called the Condon Committee after its director, physicist Edward U. Condon, the study was, nonetheless, another setup. Condon was not open to the idea of UFOs, and he fired investigators who disagreed with him. Word eventually got out (through a book by one of those fired investigators and a *Look* magazine article) that the committee's efforts were every bit as poor and deceptive as had been those of the air force before it.

The Condon Committee published its report, *Scientific Study of Unidentified Flying Objects,* in 1969. Not surprisingly, it concluded that "further scientific study of UFOs probably cannot be justified in the expectation that science will be advanced thereby." Still, the report admitted that fully one-third of its cases were unexplainable, even after in-depth study. Just as with *Project Blue Book Special Report 14,* the report's conclusion did not follow from the data inside it. Still, the air force had what it wanted: an excuse to shut down Project Blue Book, which was officially terminated on December 17, 1969.

Hollywood director Steven Spielberg consulted Josef Allen Hynek and the Center for UFO Studies when filming his 1977 motion picture *Close Encounters of the Third Kind.*

Types of UFO Sightings

UFO sightings have been reported worldwide and vary little from one nation to another. The most commonly reported UFO shapes are discs and cigars; in recent years growing numbers of reports have

A man in Kansas in 1952 saw a large disc-shaped structure like the one pictured here hovering nearby— one of the many cases Project Blue Book could not explain.

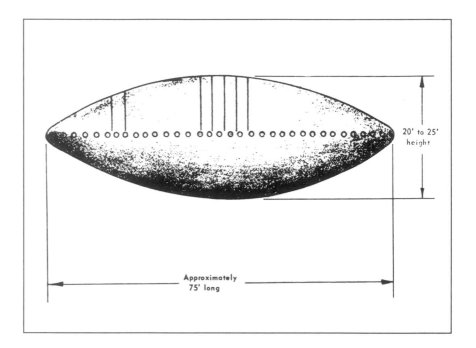

20' to 25' height

Approximately 75' long

included boomerangs and triangles. Quite a few sightings are simply points of light in the night sky. These cases have often been given common explanations—Venus, meteors, passing aircraft—but sometimes the behavior of the lights makes such explanations difficult to accept.

In his 1972 book *The UFO Experience,* Hynek divided reports into these general categories: lights seen at night; daylight discs; radar/visual cases; close encounters of the first kind (CE1s—a UFO seen at less than 500 feet from the witness); close encounters of the second kind (CE2s—a UFO that physically affects its surroundings); and close encounters of the third kind (CE3s—beings observed in connection with a UFO sighting).

The best evidence for the existence of UFOs comes from radar/visual cases and CE2s. An example of the first type occurred over several hours between August 13 and 14, 1956, at two English bases run jointly by the Royal Air Force and the U.S. Air Force. Unidentified objects traveling at high speed were tracked on air and ground radar and seen by observers on earth and pilots in flight. But the best-documented CE2 case, suggesting a UFO landing, took place late on the afternoon of January 8, 1981, in Trans-en-Provence, France. An old man working in his garden reported that he had seen the landing of a "ship ... in the form of two saucers upside down, one against the other." The object rested on the ground for a short time before flying away.

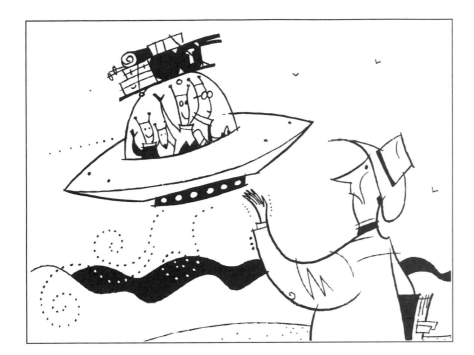

In 1959 three dozen witnesses waved at humanlike figures in a hovering UFO—and they waved back!

The site of the landing revealed traces, impressions, and other evidence that a large vehicle had been there. France's official UFO-investigating agency, Group d'Etude des Phenomenes Aerospatiaux Non-Identifies (GEPAN), began an intense study, taking soil, leaf, and plant samples to France's leading botanical laboratory. In 1983 GEPAN released a 66-page report on the case, which noted that the leaves had mysteriously lost 30 to 50 percent of their chlorophyll and had aged quickly in a manner that could not be repeated in the laboratory. The study concluded that a combination of "mass, mechanics, a heating effect, and perhaps certain transformations and deposits of trace minerals [phosphate and zinc]" had altered the site, leading scientists to believe "that something similar to what the eyewitness has described did take place there."

CE3s are usually the strangest UFO claims and the ones most likely to cause sensational publicity. And for many ufologists, they are also the most difficult to accept. In most cases the witnesses—both separately and in groups—seem believable, and psychological testing shows that they are not mentally ill. CE3s range from brief sightings of humanoids (beings humanlike in appearance; nearly all CE3s describe humanoids) to abductions, where observers are taken against their will into UFOs and aliens perform strange experiments on them.

Phony photographs, such as this "flying cup" taken during the 1950s, are the most common type of UFO hoax.

One of the most spectacular CE3 reports was logged at Boianai, Papua New Guinea, on the evenings of June 26 and 27, 1959. An Anglican missionary from Australia, the Reverend W. B. Gill, and three dozen other witnesses observed some well-lit human-like figures through the dome of a hovering UFO. Gill thought they were "busy at some unknown task." During the second sighting, he and the others waved at the figures, who waved back!

UFO Hoaxes and Contacts

Many people believe that UFO reports are tricks or hoaxes. In truth, the majority of UFO sightings are honest mistakes—misidentifications—and hoaxes are actually rare. Even the air force has found that only about 1 percent of the reports it receives involve trickery, with many of these centering on UFO photographs, which are very easy to fake. Still, hoaxes have been recorded.

For instance, just a few days after Kenneth Arnold's "flying saucer" sighting in 1947, two men in Tacoma, Washington, displayed some melted metal remains that they claimed had dropped from a "flying doughnut" hovering above nearby Maury Island. While investigating the claim, two army air force officers were killed in a plane crash, leading to rumors that they had been murdered for knowing too much. The men's story, however, ended up being nothing more than a practical joke that had gotten way out of hand.

From the early 1950s on, colorful figures, based mostly in southern California, claimed ongoing contact with kind visitors from Venus, Mars, Saturn, and other planets. Many of these "contactees" also told of space travel and meetings with extraterrestrials or "space brothers" in other worlds (also see entries: **Space Brothers** and **Ummo**). As proof they offered suspiciously clear close-up photographs of spaceships and conveniently blurred photographs of their space-brother friends.

The most famous contactee was George Adamski, whose adventures began on November 20, 1952, when he reportedly met Orthon, a visitor from Venus, in the California desert. Others would also claim contact and write books and lecture about their experiences, attracting followers who were interested in the most fantastic—or occult—features of UFO experiences. Although claims by contactees were often exposed as embarrassing lies, the belief of their fans remained unshakable.

In fact, most of the contactees were not tricksters; many believed that they were in psychic or spiritual—if not physical—contact with space people. Psychic contactees did not feel pressured to produce "evidence" of their meetings, though they did indicate their belief in strong, and even shocking, ways. Gloria Lee, for example, fasted to death after a friend from Jupiter told her to do so. And, through automatic writing (writing performed without thinking, seemingly directed by an outside force), Dorothy Martin of Oak Park, Illinois, received messages from the spaceman Sananda, who warned her of dire geological disasters to occur on December 20, 1954. She and her followers alerted the press, quit their jobs, and planned to escape by spaceship on the dreaded date. When the saucer failed to show, Martin and her group were ridiculed the world over.

Theories about UFOs

Until the mid-1960s, there were two primary explanations regarding UFOs. One was that UFOs were hoaxes or mistaken identifications. The other was that they were spacecraft from outside our world. Donald H. Menzel, a Harvard University astronomer, was a leading supporter of the first school of thought, and Donald E. Keyhoe, an aircraft writer, was a leading supporter of the second. Both wrote books and articles arguing their positions and, in turn, gained powerful supporters in science, government, and the military.

Near the end of the decade, some ufologists began considering new explanations for UFO accounts. They began to think that the key to the mystery lay in the strangest reports, which traditional ufologists—concerned with believability, documentation, and evidence—had often laughed at or ignored. Some ufologists began to think that UFO contact stories did not really involve flesh and blood visitors from other planets but that they were visions that sprang from a witness's imagination. Perhaps contact experiences were unusually vivid dreams; perhaps abduction by extraterrestrials was just a Space Age version of the "kidnapped by fairies" story that people told in earlier ages. This "psychosocial hypothesis" concerning UFO experiences became a major force in the study of UFOs, especially in Europe.

REEL LIFE

The Blob, 1958.

Sci-fi thriller about a small town's fight against a slimy Jell-o invader from space. Slightly rebellious Steve McQueen (in his first starring role) redeems himself when he saves the town with quick action. This low-budget horror/teen-fantasy became a camp classic. A hi-tech remake of the same name made in 1988 is an excellent tribute to this film.

E.T.: The Extra-Terrestrial, 1982.

The story of an alien creature stranded on earth and his special bonding relationship with a young boy. A modern fairy tale providing warmth, humor, and sheer wonder, this is one of the most popular films in history.

It Came from Outer Space, 1953.

Aliens take on the form of local humans to repair their spacecraft in a small Arizona town. A fine science-fiction film based on a story by sci-fi writer Ray Bradbury.

Crashes and Cover-ups

The psychosocial approach to UFO study did not take lasting hold in the United States, however. One reason for this was the release, by the late 1970s, of many formerly secret government UFO reports through the Freedom of Information Act. Many outstanding radar/visual cases and other sightings were uncovered, exciting traditional ufologists. These discoveries also renewed suspicions that the government was involved in UFO cover-ups.

Keyhoe and others who suspected a cover-up thought that the air force might be hiding other cases of radar trackings, films, and even testimonies of pilots who had managed contact with UFOs. Some thought that the air force might be hiding even stronger proof of visitors from space, such as the remains of crashed saucers and the bodies of their pilots. While no evidence existed to support these suspicions, the stories would not go away. In the 1970s ufologist Leonard H. Stringfield started collecting reports and interviewing people who claimed knowledge, sometimes firsthand, of such evidence.

Two other ufologists, Stanton T. Friedman and William L. Moore, focused on one particular occurrence, the supposed crash of a UFO in Lincoln County, New Mexico, in early July 1947 (also see entry: **Hangar 18**). They interviewed nearly three dozen people who were directly involved and also spoke with another 50 who had indirect involvement. A few years later a Chicago organization, the J. Allen Hynek Center for UFO Studies (CUFOS), conducted its own study, bringing the total of sources, ranging from area ranchers to air force generals, to over four hundred!

Called "the Roswell incident" (after Roswell Field, New Mexico, which the air force used as its first base of investigation), the case was well documented and truly puzzling. Yet as would be expected, Friedman and Moore sometimes met with people whose claims sounded too fantastic. Some related tales not only of spaceship crashes but of face-

Budd Hopkins

(1931-)

Budd Hopkins, whose specialty is UFO-abduction reports, first became interested in UFOs when he and two companions observed a disc in the sky in 1964. It was not until a friend told him about witnessing a UFO landing, complete with occupants, that Hopkins became active in the field.

With the help of psychiatrist Robert Naiman and psychologist Aphrodite Clamar, Hopkins tried to counter the memory loss that many UFO witnesses experienced after being abducted. Under hypnosis, some of these people told of frightening meetings with large-headed, gray-skinned humanoids who forced them into medical examinations. Hopkins published his first book of research, the popular *Missing Time,* in 1981.

In his second book, *Intruders* (1987), Hopkins updated his abduction research but presented findings that were more fantastic. A frequent lecturer on the subject, Hopkins created the Intruders Foundation in New York City in 1990 to fund research and to offer therapy to witnesses haunted by their disturbing memories.

to-face contact between aliens and U.S. government officials. One investigator close to the incident reported that some military and intelligence insiders promised him a "truckload of documents" to support their incredible stories, but they ultimately produced only a handful of papers. The most disturbing of these was a document that arrived one day in December 1984 in an envelope with no return address.

Inside the envelope was a roll of 35-mm film that, when developed, showed a portion of a presidential briefing document dated November 18, 1952. It appeared to have been written by Vice Admiral Roscoe H.

Hillenkoetter, telling then-President-elect Dwight D. Eisenhower of two UFO crashes: one in Roswell in 1947, the other along the Texas-Mexico border in 1950. It also spoke of "Operation Majestic-12," a scientific, military, and intelligence force set up to study the wreckage and bodies of the space beings (called "extraterrestrial biological entities" or EBEs).

When the document copy was released to the press, it caused a great uproar and massive publicity, including coverage in the *New York Times* and on television news show *Nightline*. The FBI launched an investigation, but it had as little luck as ufologists in getting to the bottom of the matter. Because the signature on the document appeared suspicious (as did its format), the copy was eventually judged a forgery. Why the unknown forger carried out his hoax remains a mystery.

The Future of Ufology

In recent years more social scientists and mental health professionals have become interested in UFO study. They are especially drawn to stories of UFO abductions reported by seemingly normal people. These professionals are eager to find out if such experiences spring from within an individual or really do come from outside the physical world.

Sources:

Clark, Jerome, *The Emergence of a Phenomenon: UFOs from the Beginning through 1959—The UFO Encyclopedia,* Volume 2, Detroit, Michigan: Omnigraphics, 1992.

Clark, Jerome, *UFOs in the 1980s: The UFO Encyclopedia,* Volume 1, Detroit, Michigan: Apogee Books, 1990.

Fawcett, Lawrence, and Barry J. Greenwood, *Clear Intent: The Government Coverup of the UFO Experience,* Englewood Cliffs, New Jersey: Prentice-Hall, 1984.

Hendry, Allan, *The UFO Handbook: A Guide to Investigating, Evaluating and Reporting UFO Sightings,* Garden City, New York: Doubleday and Company, 1979.

Hopkins, Budd, *Intruders: The Incredible Visitations at Copley Woods,* New York: Random House, 1987.

Hynek, J. Allen, *The UFO Experience: A Scientific Inquiry,* Chicago: Henry Regnery Company, 1972.

Jacobs, David Michael, *Secret Life: Firsthand Accounts of UFO Abductions,* New York: Simon and Schuster, 1992.

Randle, Kevin D., and Donald R. Schmitt, *UFO Crash at Roswell,* New York: Avon Books, 1991.

UNIDENTIFIED AIRSHIPS

Reports of unidentified airships began before the turn of the twentieth century, preceding and foreshadowing the phenomena of unidentified flying object reports that began to flourish in the 1940s. The first known printed account of a mysterious "airship" appeared in the March 29, 1880, issue of the *Santa Fe Weekly New Mexican*. The newspaper reported that late on the evening of March 26, observers in the village of Galisteo Junction watched a "large balloon" pass overhead and heard the merry shouts of its passengers. A couple of odd objects were dropped from the craft: a cup of "very peculiar workmanship" and a "magnificent flower, with a slip of exceedingly fine silk-like paper, on which were some characters resembling those on Japanese tea chests." The next evening a Chinese American visitor said that the paper carried a message from his girlfriend. According to the story she was a passenger on the airship, which was headed for New York City.

American papers back then tended to treat airship sightings as jokes—and were, in fact, behind many of the accounts themselves.

Like many other airship stories reported in the late nineteenth-century press, this one is almost certainly a tall tale. American papers back then tended to treat airship sightings as jokes—and were, in fact, behind many of the accounts themselves. In the years ahead, however, more believable reports would be made in the United States and other countries. And it is likely that these strange airships would have been viewed as UFOs (unidentified flying objects) had the sightings occurred decades later, in the second half of the twentieth century. In fact, sightings of airshiplike objects—cigar-shaped, with multicolored lights along the sides and flashing searchlights—continue to this day.

An outbreak of airship reports occurred along the border of Germany and Russian Poland in early 1892. As would be the case with later airship scares, the Germans were thought to have developed advanced aircraft that could fly against the wind (unlike balloons) and hover for long periods. No such aircraft existed at the time, nor had any been developed by 1896, when the great American airship scare hit California.

California Airship Scare

Beginning in mid-November 1896, many witnesses in both city and country areas of California reported seeing fast-moving or still lights at night and assumed that they were connected to airships. A daylight sighting reported in the *San Francisco Call* of November 22 described a "balloon" traveling on end, "with what appeared to be

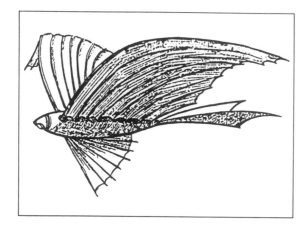

Dallas Morning News, 1897, sketch of an airship reported over Texas.

wings both before and behind the [bottom] light." A "great black cigar with a fishlike tail ... at least 100 feet long" and a surface that "looked as if it were made of aluminum" was described in another account in the December 1 issue of the *Oakland Tribune.* In other cases observers reported airships with propellers.

All during this time, the press was focusing its attention on George D. Collins, a San Francisco lawyer who claimed that he not only represented the inventor who created the airship, but had even seen the wonderful craft himself (both claims he later denied). The inventor was rumored to be E. H. Benjamin, a dentist from Maine who liked to tinker with machines. While Benjamin insisted that his "inventions [had] to do with dentistry," some refused to believe him and bothered the man so much that he had to go into hiding. Reporters even broke into his office in search of evidence—but found nothing but dental filings!

Then, according to an article in the *Oakland Tribune,* former California attorney general W. H. H. Hart claimed to represent the inventor of the mysterious airship. He said that Collins had been fired for talking too much. But Hart proved to be even more gabby, stating that there were actually two airships and that they would be used to bomb the Spanish fort in Havana, Cuba. When pressed for proof, Hart—like Collins before him—backed down, admitting that he had not personally seen the invention and had only met with someone who claimed to be the inventor.

The California airship scare soon faded away. But in February 1897, Nebraska newspapers began reporting night sightings in country districts of lights moving with "most remarkable speed." On February 4 witnesses at Inavale got a close look at the object to which the lights were attached: it was cone-shaped, 30 to 40 feet long, and had "two sets of wings on a side, with a large, fan-shaped rudder," according to the *Omaha Daily Bee* (February 6). Over the following weeks, a flurry of sightings were reported in Nebraska and then in neighboring Kansas. By early April airships were moving east, north, and south, and all that month newspapers were filled with sightings, rumors, and tall tales.

More Hoaxes

Many of these questionable stories focused, as they had in California, on secret inventors. Some accounts even reported that airships had landed and that their passengers, ordinary Americans, had identified themselves and spoken of their plans. These "conversations" with air travelers would appear word for word in newspaper stories. While they were treated like serious news, these accounts were almost certainly supplied by imaginative writers!

Other hoaxes focused on the idea that mysterious airships came from outer space. A rancher from Le Roy, Kansas, swore that he, his son, and his hired man had seen strange-looking beings in an airship lasso and steal a calf from a corral outside his house. Though the tale attracted wide attention (and was rediscovered and widely published in UFO literature of the 1960s), it turned out to be a prank played by the rancher and fellow members of a local liars' club.

One rancher swore that strange-looking beings in an airship roped and stole a calf from his corral.

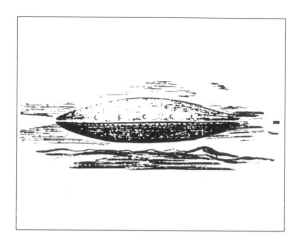

Although there were no blimps in the United States in the late 1890s, there were many reported sightings of flying cigar-shaped structures.

Similarly, the *Dallas Morning News* of April 19, 1897, printed an Aurora, Texas, man's report of a local airship crash—and of the burial of its only passenger, a Martian, in the local cemetery. Invented as a joke, the story was rediscovered in the 1960s and 1970s and brought several shovel-carrying searchers to the tiny, fading village.

Amid all the hoaxes, though, there *were* real reports of cigar-shaped structures, with or without wings, and of night sightings of lights. Perhaps behind all the tall tales and the silliness, the first great modern UFO wave—with a full variety of UFO types—was taking place.

Twentieth-Century Sightings

While by late May 1897 the flurry of sightings had quieted, reports of airships continued into the next century. In the summer of 1900, for example, two young men from Reedsburg, Wisconsin, saw a huge blimp-shaped structure hovering in the night sky. As it passed over a grove of trees, the trees bent as if blown by a strong wind, though the night was still. The March 15, 1901, issue of New Mexico's *Silver City Enterprise* even reported that a local doctor had taken a clear photograph of an airship. But the picture was lost.

In 1901 a wave of airship sightings occurred in Great Britain, the United States, New Zealand, and Australia. In Britain the sightings began in March, and most described torpedo-shaped craft moving at a "tremendous pace" and flashing lights and searchlights; these renewed fears—first expressed 15 years earlier in eastern Europe and still unsupported—of high-flying German spies. In America, secret inventors were suspected in this new flurry of airship sightings.

New Zealand's wave began in July 1901 at the southern end of South Island and then moved northward. As with other airship scares, some witnesses there claimed to have seen humanlike figures in passing craft. In one case, said to have taken place on August 3, a Waipawa man reported that an airship passenger had shouted at him in an unknown language. In another account, a man in a boat thought he was being attacked when "missiles" were fired from an airship and hit the water. Australia also experienced a handful of sightings that August.

Another airship-sighting wave occurred in the fall of 1912, with reports all across Europe. Most of the objects were described as large and cigar-shaped, with very bright searchlights. Few, if any, of the accounts mentioned wings. And, as before, the airships could hover and move at great speeds, even against the wind. While this wave of sightings died down by the following April, airship reports continued from time to time in Europe and elsewhere.

On October 10, 1914, for example, a Manchester, England, man claimed that he saw an "absolutely black, spindle-shaped object" cross the face of the sun. A cigar-shaped object at least 100 feet long flew over Rich Field, Waco, Texas, one evening in early 1918. It left witnesses with—in the words of one— "the weirdest feeling of our lives." And in the summer of 1927, an airship was seen over Wolfe County, Kentucky. One observer compared it to a "perfectly shaped, huge fish, with big fins extended outward near the front and small, short ones near the rear."

After the 1920s, unidentified cigar-shaped objects were rarely called "airships." Still, they continued to be reported. On October 9, 1946, observers in San Diego, California, saw an airshiplike object that they compared to a "huge bat with wings." A similar object was seen over Havana, Cuba, the following February.

Driving to work at 5:50 A.M. on August 25, 1952, a Pittsburg, Kansas, radio musician said he came upon a 75-foot-long object with windows through which the head and shoulders of a human figure were visible. He told air force investigators that along the UFO's outer edges "were a series of propellers about six inches to eight inches in diameter, spaced closely together." And on the morning of February 6, 1967, Ruth Ford sighted two fast-moving "cigar-shaped craft"—each with two small propellers and a row of windows—as she drove between Deming and Las Cruces, New Mexico. She could see no one inside.

In the summer of 1927, an airship was seen over Wolfe County, Kentucky. One observer compared it to a "perfectly shaped, huge fish, with big fins extended outward near the front and small, short ones near the rear."

Sources:

UFOs from the Beginning through 1959—The UFO Encyclopedia, Volume 2, Detroit, Michigan: Omnigraphics, 1992.

Cohen, Daniel, *The Great Airship Mystery: A UFO of the 1890s,* New York: Dodd, Mead and Company, 1981.

Lore, Gordon I. R., Jr., and Harold H. Deneault, Jr., *Mysteries of the Skies: UFOs in Perspective,* Englewood Cliffs, New Jersey: Prentice-Hall, 1968.

Ancient ETs and Their Calling Cards

- ANCIENT ASTRONAUTS
- NAZCA LINES
- SIRIUS MYSTERY

Ancient ETs and
Their Calling Cards

ANCIENT ASTRONAUTS

In the 1970s, the idea that advanced space beings had visited the Earth early in man's history—and had played a part in the development of human intelligence and technology—became very popular. Sparked by the publication of Swiss writer Erich von Däniken's wildly popular book *Chariots of the Gods?: Unsolved Mysteries of the Past* in 1968, this "ancient astronaut" movement swept across Europe and then into Great Britain and the United States.

Von Däniken and other believers of the theory argued that space beings were behind the archaeological and engineering wonders of the ancient world, like the Egyptian pyramids and Peru's **Nazca lines.** They also believed that the gods of Judaism, Christianity, and other religions were actually extraterrestrials (beings from other planets) who, by mating with our primitive ancestors—or by changing their genes—created *Homo sapiens,* or modern men and women. According to von Däniken and his followers, God himself was an astronaut.

Von Däniken had no great interest in ufology, the study of unidentified flying objects, nor did he have scientific training. But at age 19 he did have a mystical vision that led him to "the firm belief that the earth had been visited by extraterrestrial astronauts." He began to read widely then, looking for evidence of ancient astronauts in historical and archaeological literature, even visiting archaeological sites in North Africa and the Americas. He also read the works of others who suggested that early space visits had occurred, including Jacques Bergier and Louis Pauwels's 1960 best-seller *The Morning of the Magicians* and Robert Charroux's 1963 book *One Hundred Thousand Years*

According to von Däniken and his followers, God himself was an astronaut.

Erich von Däniken.

of Man's Unknown History. Von Däniken would freely borrow ideas from both of these works.

Indeed, the theory of ancient astronauts was nothing new. It had been around for quite a long time, in fact. In the late nineteenth century, followers of the theosophy—teachings about God and the world based on mystical insight—of Helena Petrovna Blavatsky believed that space people played a part in human history. Many writers addressed the subject before von Däniken, including flying-saucer contactee George Hunt Williamson, who produced three books on ancient-astronaut themes in the 1950s. And M. K. Jessup (also see entry: **The Philadelphia Experiment**), a former astronomer, wrote *The Case for the UFO* (1955), *UFO and the Bible* (1956), and *The Expanding Case for the UFO* (1957), all detailing past and present alien influences. To Jessup, however, the "aliens" were earthlings: pygmy races who tens of thousands of years ago developed antigravity technology and escaped to the moon and beyond just as natural disasters wiped out other advanced civilizations. According to Jessup, this population nonetheless continues to observe Earth and may even be the "little people" so often described in folklore and UFO reports.

Clearly, von Däniken's ideas were not original, but he put them together in such a way and at the precise time that they would cause a sensation. While most scientists and journalists strongly questioned his theories and his "evidence"—based more on personal judgments than solid proof and careful reasoning—the public couldn't seem to get enough of the subject. When *Chariot of the Gods?* was made into a film in Germany, audiences flocked to the box office, and when an edited version was aired on television in the United States, 250,000 copies of the paperback were sold in the following 48 hours! Von Däniken wrote a number of successful sequels to his best-seller, triggering a flood of similar books by other authors.

An Ancient Astronaut Society was formed in 1973, headed by attorney Gene M. Phillips in the United States and von Däniken in his native Switzerland. Sponsoring archaeological expeditions to sites where "members may have an opportunity to examine the evidence firsthand," the organization was created—according to Phillips—to "*prove* that civilization, technology, and intelligence *originated* in outer space." Although by the early 1980s the ancient astronaut fad had run its course, the society remains active and publishes a bimonthly bulletin, *Ancient Skies.*

Sources:

Story, Ronald, *The Space-Gods Revealed: A Close Look at the Theories of Erich von Däniken,* New York: Harper and Row, 1976.

Von Däniken, Erich, *Chariots of the Gods?: Unsolved Mysteries of the Past,* New York: G. P. Putnam's Sons, 1969.

NAZCA LINES

At some time before 1000 B.C., the Nazca Valley, a desert region on Peru's southern coast, was inhabited by a people who developed advanced farming methods that allowed them to build an irrigation system, improve their crops, and expand the area of land they could farm. Over the next 1,500 years, they also developed outstanding skills in weaving, pottery, and architecture. Yet perhaps the most fascinating of their cultural achievements was the creation of a remarkable ground art—the exact purpose of which remains a mystery.

The so-called Nazca lines, of which there are thousands, consist, according to investigator William H. Isbell, of five kinds of markings: long straight lines; large geometric figures; drawings of plants and animals; rock piles; and figures decorating hillsides.

The lines may be as narrow as six inches or as wide as several hundred yards. Some run for many miles. The Nazca people created some of them by removing dark surface stones and placing them in the desired patterns. For others, according to William E. Shawcross, they removed the desert's "thin brown surface coating" by walking or sweeping across it, "[exposing] the creamy pink soil underneath." Because of the area's dry, stable climate, these light-colored Nazca lines have remained nearly unchanged for many centuries.

Nazca lines decorate hillside.

What makes the Nazca markings so very odd, though, is the fact that a great many of the forms are discernible only from the air! Archaeologists have developed several explanations for this: one is that the figures, probably of religious significance, were not meant to be seen as a whole by human eyes; a second is that the Nazca people built balloons that allowed them to view the figures when they flew over the sites. This suggestion, while not impossible, lacks supporting evidence.

Signals for the Gods?

The Nazca lines attracted public attention not long after the heyday of UFO sightings began. In the 1950s, as more and more books and magazine articles addressed UFOs, some writers looked back to ancient history and mythology for evidence of early space visitors. In

Nazca lines viewed from the air—who or what could have viewed them from above?

an article in the October 1955 issue of *Fate,* James W. Moseley suggested that since the markings were largely invisible from the ground, the Nazca people must have "constructed their huge markings as signals to interplanetary visitors or to some advanced earth race ... that occasionally visited them."

In his 1959 book *Road in the Sky,* flying-saucer "contactee" George Hunt Williamson included a whole chapter on the mysterious Nazca lines. Like Moseley, Williamson believed that "sky gods" or space beings visited Earth in the distant past and that the Nazca lines were connected to them. Williamson also wondered if these space visitors were somehow related to the advanced civilizations, like Lemuria and Atlantis, described in ancient myths. He thought that perhaps the Nazca lines and other puzzling archaeological sites served as "magnetic centers," locations at which spaceships could refuel.

In the early 1960s a French best-seller by Louis Pauwels and Jacques Bergier, published in America as *The Morning of the Magicians,* included the Nazca lines in its theory of **ancient astronauts.** This idea, that advanced space beings visited the earth early in man's history and played a part in the development of human intelligence and technology, reached its greatest popularity with Swiss writer Erich von Däniken's book *Chariot of the Gods?,* first published in West Germany in 1968 and reprinted in translated editions around the world. According to von Däniken, the Nazca lines marked out an "airfield" on which spacecraft landed and took off.

Still, nothing in the nature of these lines points to such a purpose. In fact, a critic of von Däniken's ideas stated, "It hardly seems reasonable that advanced extraterrestrial spacecraft would require *landing strips,*" adding that Nazca's "soft, sandy soil" was hardly suitable for an airport. Regardless, this explanation of the Nazca lines was accepted for a time during the ancient astronaut craze of the 1970s.

Sources:

Pauwels, Louis, and Jacques Bergier, *The Dawn of Magic,* London: Anthony Gibbs and Phillips, 1963.

Story, Ronald, *The Space-Gods Revealed: A Close Look at the Theories of Erich von Däniken,* New York: Harper and Row, 1976.

Von Däniken, Erich, *Chariots of the Gods?: Unsolved Mysteries of the Past,* New York: G. P. Putnam's Sons, 1970.

SIRIUS MYSTERY

The **ancient astronaut** fad of the 1970s gave rise to a number of questionable books on the subject, works that were weak on evidence and careful reasoning. An exception to this was Robert K. G. Temple's *The Sirius Mystery* (1977), a thoughtful, well-researched look at the possible early influence of space beings on the Dogon, a tribe in West Africa.

The Dogon are believed to be of Egyptian descent. After living in Libya for a time, they settled in Mali, West Africa, bringing with them astronomy legends dating from before 3200 B.C. In the late 1940s four of their priests told two French anthropologists of a secret Dogon myth about the star Sirius (8.6 light-years from the earth). The priests said that Sirius had a companion star that was invisible to the human eye. They also stated that the star moved in a 50-year elliptical orbit around Sirius, that it was small and incredibly heavy, and that it rotated on its axis.

All of these things happen to be true. But what makes this so remarkable is that Sirius's companion star, called Sirius B, was first photographed in 1970. While people began to suspect its existence around 1844, it was not seen through a telescope until 1862—and even then its great density was not known or understood until the early decades of the twentieth century. The Dogon beliefs, on the other hand, were supposedly thousands of years old!

Even if these people had somehow seen Western astronomy textbooks, they could not have known about Sirius B. Also puzzling was their knowledge of the rotations and orbits of planets in our solar system and of the four major moons of Jupiter and the rings of Saturn. How did they learn all this? Dogon folklore says that this knowledge came from unearthly sources.

The Dogon tell the legend of the Nommos, awful-looking beings who arrived in a vessel along with fire and thunder. The Nommos, who could live on land but dwelled mostly in the sea, were part fish, like **merfolk** (mermaids and mermen). Similar creatures have been noted in other ancient civilizations—Babylonia's Oannes, Acadia's Ea, Sumer's Enki, and Egypt's goddess Isis. It was from the Nommos that the Dogon claimed their knowledge of the heavens.

The Dogon also claimed that a third star existed in the Sirius system. Larger and lighter than Sirius B, this star revolved around Sirius as well. And around it orbited a planet from which the Nommos came.

> Also puzzling was their knowledge of the rotations and orbits of planets in our solar system and of the four major moons of Jupiter and the rings of Saturn.

The Nommos, who were part fish, came to Earth in a strange vessel accompanied by fire and thunder.

Other Explanations

Although Temple's *The Sirius Mystery* was taken more seriously than many other ancient-astronaut writings when it was first published, it met with some bad luck; it was criticized by two important science figures, writer Ian Ridpath and celebrity-astronomer Carl Sagan. From that point on many felt that it did not get the kind of consideration that its well-laid-out case deserved.

Ridpath and Sagan had their own simple explanation for the Sirius mystery: the Dogon got their supposedly ancient knowledge of the heavens from modern informants. They asserted that Westerners had probably discussed astronomy with Dogon priests, who quickly added this new information to older folklore. French anthropologist Germaine Dieterlen, who had lived among the Dogon for most of her life and whose writings on their astronomy myths had caught Temple's attention in the first place, called this idea "absurd" when asked about it by a reporter for BBC-TV's *Horizon* program. Then she displayed for the show's audience a Dogon object crafted 400 years ago, which clearly indicated Sirius and its companion stars.

Despite the criticism, Temple continued to defend his position. He pointed out that some of the information, like that concerning the super-weight of Sirius B, was only a few years old when anthropologists first collected it from the Dogon in 1931. Temple wondered why Western astronomers would rush to far-off Mali to share their new astronomical knowledge with Dogon priests? And how, in two or three years' time, could this information then filter down through the entire Dogon and surrounding cultures of over two million people and show up in hundreds of thousands of objects, woven blankets, carved statues, and more?

These reasonable questions brought no response from Ridpath, Sagan, and other Temple critics. Whenever a writer of an article or book would report that the Dogon's Sirius beliefs came from modern informants, Temple would respond with a point-by-point account that argued otherwise. But his comments were simply ignored.

Now there are other factors that cast serious doubt on the Dogon's story. So far no third star has been detected in the Sirius system. And for scientists who search for evidence of intelligent life in the universe, Sirius has never been on their list of places to look.

Still, Temple raised serious questions about the Dogon's Sirius beliefs. Almost two decades after the publication of *The Sirius Mystery,* the book has been nearly forgotten. Yet the puzzle of the Dogon's remarkable astronomical knowledge remains.

Sources:

Ridpath, Ian, *Messages from the Stars,* New York: Harper and Row, 1978.
Sagan, Carl, *Broca's Brain: Reflections on the Romance of Science,* New York: Random House, 1979.
Story, Ronald, *Guardians of the Universe?,* New York: New English Library, 1980.
Temple, Robert K. G., *The Sirius Mystery,* New York: St. Martin's Press, 1977.

Unexpected Guests and Interplanetary Communications

- **FLYING HUMANOIDS**

- **HAIRY DWARFS**

- **FLATWOODS MONSTER**

- **SPACE BROTHERS**

- **MEN IN BLACK**

- **UMMO**

Unexpected Guests and Interplanetary Communications

FLYING HUMANOIDS

When a mysterious object passed over Mount Vernon, Illinois, on the evening of April 14, 1897, one hundred citizens, including Mayor B. C. Wells, saw something that "resembled the body of a huge man swimming through the air with an electric light on his back."

Batman and Birdman

Though rare, accounts like this one—of flying beings of human appearance—do occur from time to time. A "winged human form," for example, was observed over Brooklyn, New York, on September 18, 1877, according to W. H. Smith in the *New York Sun*. Three years later the *New York Times* of September 12, 1880, recorded reports from Coney Island of a "man with bat's wings and improved frog's legs ... at least a thousand feet in the air ... flying toward the New Jersey coast ... [with] a cruel and determined expression."

V. K. Arsenyev, a Russian writer, reported seeing a mysterious flying creature in the Sikhote Mountains near Vladivostok, Russia, on July 11, 1908. In a 1947 book he recalled the experience:

> The rain stopped, the temperature of the air remained low and the mist appeared over the water. It was then that I saw the mark on the path that was very similar to a man's footprint. My dog Alpha bristled up, snarled, and then something rushed about nearby trampling among the bushes. However, it didn't go away.

Arsenyev related that when he threw a stone "towards the unknown animal ... something happened that was quite unexpected": he "heard the beating of wings. Something large and dark emerged from the fog and flew over the river. A moment later it disappeared in the dense mist." When the writer told some local men about his experience, "they broke into a vivid story about a man who could fly in the air. Hunters often saw his tracks, tracks that appeared suddenly and vanished suddenly in a way that could only result if the 'man' alighted on the ground, then took off again into the air."

One night in 1952, U.S. Air Force Private Sinclair Taylor, on guard duty at Camp Okubo, Kyoto, Japan, said he heard a loud flapping noise. Looking up, he saw a huge "bird" in the moonlight. When it approached, he became frightened and put a round of ammunition into his gun. The "bird" now had stopped its flight and was hovering close by, staring at the soldier.

"The thing, which now had started to descend again, had the body of a man," Taylor recalled. "It was well over seven feet from head to feet, and its wingspread was almost equal to its height. I started to fire and emptied my carbine where the thing hit the ground. But when I looked ... to see if my bullets had found home there was nothing there." When the sergeant of the guard came to investigate the gunshots and heard the story, he told Taylor that he believed him—because a year earlier another guard had seen the same thing!

Another soldier's tale of a flying humanoid was reported by Earl Morrison, who served with the First Marine Division in Vietnam. While stationed near Da Nang in August 1969, he and two other guards saw an extraordinary sight just after one o'clock in the morning. They were sitting atop a bunker and talking when they noticed something approaching them in the sky. Morrison related:

> We saw what looked like wings, like a bat's, only it was gigantic compared to what a regular bat would be. After it got close enough so we could see what it was, it looked like a woman. A naked woman. She was black. Her skin was black, her body was black, the wings were black, everything was black. But it glowed. It glowed in the night—kind of a greenish cast to it.

The men watched the creature move about in the sky. At one point she was right above their heads, just six or seven feet up. She moved silently, without flapping her wings. Morrison said that she

blocked the moon once, but despite the increased darkness, they could still see her because she glowed brightly. It was only when the creature started to fly away that the men heard a flapping sound.

Morrison thought the covering on her skin was more like fur than feathers. "The skin on her wings looked like it was molded on to her hands," he said, and the movement of her arms suggested they had no bones in them.

In the 1950s, a Kansas boy said he saw a dark-skinned little man with pointed nose and ears.

Flying Humanoids and UFOs

In the second half of the twentieth century, most sightings of flying humanoids have been connected with UFOs.

Not all of the sightings have been of winged figures. Sometimes the humans or humanoids fly through the air with the aid of machines attached to their bodies. The first known account of this kind took place near Louisville, Kentucky, on July 29, 1880, according to the *Louisville Courier-Journal*. Another such case occurred in Chehalis, Washington, on January 6, 1948, when an elderly woman and a group of children reported seeing a man with long mechanical wings. Flying in an upright position, he moved the wings with instruments on his chest. Six and a half years later, the *Wichita Evening Eagle* reported that a 12-year-old Coldwater, Kansas, farm boy had observed a dark-skinned little man with pointed nose and ears float toward a UFO that was hovering nearby.

Three Houston, Texas, residents reported what may or may not have been a winged UFO in the early morning hours of June 18, 1953. As they sat on the front porch of their apartment building trying to escape the heat, a huge shadow fell across the lawn, then appeared to bounce into a pecan tree. They saw the "figure of a man ... dressed in gray or black [fitted] clothes," surrounded by a "dim gray light." Witnesses could not agree on whether he was wearing a cape or had wings. After 15 minutes the figure "just melted away," and soon after, a "loud swoosh" sounded across the street and a rocket-shaped object shot up and disappeared along the horizon.

Sources:

Bord, Janet, and Colin Bord, *Alien Animals,* Harrisburg, Pennsylvania: Stackpole Books, 1981.

Clark, Jerome, and Loren Coleman, *Creatures of the Outer Edge,* New York: Warner Books, 1978.

Keel, John A., *The Mothman Prophecies,* New York: E. P. Dutton and Company, 1975.

Keel, John A., *Strange Creatures from Time and Space,* Greenwich, Connecticut: Fawcett Gold Medal, 1970.

One witness saw six hairy dwarfs hauling rocks into a nearby spaceship.

HAIRY DWARFS

During the fall of 1954 a flurry of UFO sightings occurred world-wide. Many of the reports told of humanoid UFO occupants that resembled hairy dwarfs. On October 9, for example, three children roller-skating in the French countryside reported that a "round shiny machine came down very close to us. Out of it came a kind of man, four feet tall, dressed in a black sack.... His head was hairy, and he had big eyes. He said things to us that we couldn't understand and we ran away. When we stopped and looked back, the machine was going up into the sky very fast."

Five days later a French miner came upon a humanoid with a squat, furry body and oversized, slanted, bulging eyes. It was wearing a skullcap (a close-fitting cap without a brim) and had a flat nose and thick lips.

In Venezuela in early December of that year, there were several reports of night meetings with three-foot-tall, hairy—and hostile—dwarfs. In one case, said to have taken place on December 10, four such beings stepped out of a hovering UFO and tried to kidnap a young man. His companion, who happened to be armed because the two were hunting at the time, struck one of the beings on the head with the end of his gun. The gun butt splintered as if it had met with solid rock. The two men, bruised, cut, and terrified, told their story to the police soon afterward.

Nine days later, at Valencia, a jockey on a late-night training ride said he saw six hairy dwarfs hauling rocks into a nearby UFO. When they noticed him, one fired a beam of violet light and paralyzed him, even though he was trying to run away. Police found footprints at the scene, which they described as "neither human nor animal."

Accounts of meetings with UFO occupants, of course, continue to this day. But reports of sightings involving hairy dwarfs died out quickly by the end of 1954.

Sources:

Bowen, Charles, ed., *The Humanoids: A Survey of Worldwide Reports of Landings of Unconventional Aerial Objects and Their Alleged Occupants,* Chicago: Henry Regnery Company, 1969.

Lorenzen, Coral, and Jim Lorenzen, *Encounters with UFO Occupants,* New York: Berkley Publishing, 1976.

FLATWOODS MONSTER

On September 12, 1952, three boys in the tiny West Virginia town of Flatwoods (population 300) saw a slow-moving, reddish ball sail around a hill, hover briefly, and drop behind another. A bright glow seemed to come from the other side of that hill, as if the object had landed. On their way to investigate, the boys were joined by beautician Kathleen May, her two young sons, their friend Tommy Hyer, 17-year-old Eugene Lemon, and Lemon's dog.

The dog ran ahead of the group and was briefly out of sight. Suddenly it was heard barking fiercely and, moments later, seen fleeing with its tail tucked between its legs. A foul-smelling mist covered the

> When they noticed him, one fired a beam of violet light and paralyzed him, even though he was trying to run away.

ground and caused the searchers' eyes to water. Lemon and Neil Nunley, who were leading the group, reached the hilltop first and saw a "big ball of fire" 50 feet to their right when they looked down. One of the other witnesses reported that the ball was the size of a house.

An Armless, Legless Creature with an Odd-Shaped Head

To the group's left, on the hilltop and just under the branch of an oak tree, were two small lights. May suggested that Lemon turn his flashlight on them. To everyone's great amazement, the flashlight beam revealed an awful-looking creature with a head shaped—according to several of the observers—like the "ace of spades." Inside the head was a round "window," dark except for the two lights, from which pale blue beams shone straight ahead. In the short time that they observed the creature, the group members saw nothing that looked like arms or legs.

The creature, which appeared to be over six feet tall, moved toward the witnesses; it seemed to be gliding rather than walking. Seconds later it changed direction, turning toward the glowing ball down the hill. The witnesses reported that all of this took place in a matter of seconds. When Lemon fainted, the others dragged him away as they ran from the scene.

When interviewed half an hour later by A. Lee Stewart, Jr., a reporter for the *Braxton Democrat,* most members of the group were hardly able to speak. Some were given first aid. Stewart had no doubt that they had seen something that had frightened them terribly. Still, soon afterward he was able to get Lemon to take him to the hillside, where Stewart noticed an unusual odor in the grass that bothered his nose and throat. Returning to the site alone early the following morning, the reporter found "skid marks" leading down the hill toward an area of matted grass, suggesting that a large object had rested there.

Other Sightings

This strange meeting with what the press would soon call the "Flatwoods monster" took place during a flurry of sightings of unusual flying objects in the area. One man, Bailey Frame of nearby Birch River, told of seeing a bright orange ball circling over the place where the monster was reported. The object was visible for 15 minutes before it shot toward the airport at Sutton, where it was also observed.

> The creature, which appeared to be over six feet tall, moved toward the witnesses; it seemed to be gliding rather than walking.

A representation of what was reported to be seen at Flatwoods.

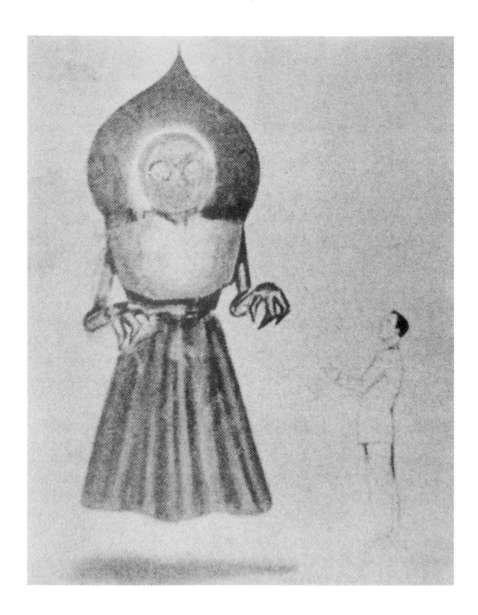

And according to another sighting, which took place a week before the Flatwoods event and 11 miles away, a Weston woman and her mother encountered a similar creature as they were driving to church. Both reported that it emitted a rank odor, and the younger woman was so frightened that she had to be hospitalized.

Of course, doubters of the Flatwoods event were plentiful; they suggested that May and her companions had seen a meteor and an owl, and only fright had caused them to think that they had seen anything else. Nonetheless, when interviewed separately shortly after the occurrence, the witnesses gave accounts that were nearly identical.

Many years later, a woman from Joliette, Quebec, would report seeing a similar creature as it gazed through a window of her home in the early morning hours of November 22, 1973. She woke her husband, who went outside to investigate. All he found was a dog that acted as if it were "scared to death." Local police believed that the woman was being truthful.

Sources:

Barker, Gray, "The Monster and the Saucer," *Fate* 6,1, January, 1953, pp. 12-17.
Sanderson, Ivan T., *Uninvited Guests: A Biologist Looks at UFOs,* New York: Cowles, 1967.

SPACE BROTHERS

On the afternoon of November 20, 1952, George Adamski—a lifelong student of the supernatural—reported meeting a being from Venus named Orthon in the southern California desert. This would be the first of many such contacts he would make with extraterrestrials: visitors from Venus, Mars, and Saturn. Adamski would also claim to travel into space, where he would attend a conference on Saturn.

Adamski related his space adventures in three books published between 1953 and 1961. The stories excited those interested in the more fantastic notions surrounding UFOs, and a movement based on these reported contacts spread from California to much of the rest of the world. Other "contactees" included Orfeo Angelucci, Truman Bethurum, Daniel Fry, Howard Menger, George Van Tassel, and George Hunt Williamson, who all published books in the 1950s and were popular lecturers in certain UFO circles.

According to these contactees, friendly, good-looking, humanlike space people come to the Earth on a peaceful mission for the Galactic Federation. For throughout the universe, the Earth is viewed as a backward place, its people primitive and violent; earthlings threaten to upset the "balance of the universe" with their nuclear weapons and warlike ways. Space people preach that if earthlings can act in a gentler manner, they will enter a New Age of peace and riches and claim their rightful place in the universe.

Early on, contactees and their followers used the affectionate nickname "space brothers" for these extraterrestrial messengers because of their kind nature and concern for the human race.

Sometimes contactees would relay this message: the Earth is about to undergo huge geological changes that will destroy much of the planet's population; those who follow the space people's direction will be saved, either by relocating to safe places, or by entering space-ships that will pick them up at the exact time of the geophysical upheaval. Early on, contactees and their followers used the affection-ate nickname "space brothers" for these extraterrestrial messengers because of their kind nature and concern for the human race.

Hoaxes

In the early years of the UFO era, it was the "physical" con-tactees—those who reported meetings with extraterrestrials and space trips—who had the greatest influence on the UFO movement. They often produced questionable photographs and other "evidence" in an effort to prove their claims. One physical contactee, Eduard

(Billy) Meier of Switzerland, was caught in a particularly embarrassing lie: his photograph of a beautiful space traveler from a "Pleiades beamship" turned out to be a fashion model whose picture had been clipped from a popular European magazine! By the early 1960s, even some of Adamski's most devoted followers had begun to doubt him as his tales grew taller and taller.

Then "psychic" contactees, those who received messages from space beings mentally, in dreams, or through automatic writing (writing performed without thinking, seemingly directed by an outside force) dominated the UFO movement. These contactees, who did not feel pressured to produce evidence, strongly believed that their extraterrestrial communications were real.

Today's average space-brothers believer might appear in Laramie, Wyoming, in the summer to attend the yearly Rocky Mountain Conference on UFO Investigation run by psychologist and contactee R. Leo Sprinkle. Most who attend live in small western towns or on farms or ranches and believe that they have been chosen to receive the mental messages of kindly beings from outside our world.

REEL LIFE

Close Encounters of the Third Kind, 1977.

Middle-American strangers become involved in the attempts of benevolent aliens to contact earthlings. This Academy Award-winning film by director Steven Spielberg is a stirring achievement. The ending is an exhilarating experience of special effects and peace-on-earth feelings.

The Day the Earth Stood Still, 1951.

A gentle alien lands on Earth to deliver a message of peace and a warning against experimenting with nuclear power. He finds his views echoed by a majority of the population, but not the ones in control. One of the truly great science-fiction films of all times.

Starman, 1984.

An alien from an advanced civilization lands in Wisconsin. Hiding in the form of a grieving young widow's recently deceased husband, he persuades the widow to drive him across country to rendezvous with his spacecraft so he can return home. Karen Allen is earthy as the widow; Jeff Bridges is fun as the likeable starman.

Sources:

Adamski, George, *Inside the Space Ships,* New York: Abelard-Schuman, 1955.

Clark, Jerome, *The Emergence of a Phenomenon: UFOs from the Beginning through 1959—The UFO Encyclopedia,* Volume 2, Detroit, Michigan: Omnigraphics, 1992.

Clark, Jerome, *UFOs in the 1980s: The UFO Encyclopedia,* Volume 1, Detroit, Michigan: Apogee Books, 1990.

Evans, Hilary, *Gods, Spirits, Cosmic Guardians: A Comparative Study of the Encounter Experience,* Wellingborough, Northamptonshire, England: The Aquarian Press, 1987.

Albert K. Bender's sketch of one of the three "Men in Black" who visited his home in 1953 and gave him the solution to the UFO mystery.

MEN IN BLACK

In 1953 Albert K. Bender of Bridgeport, Connecticut, suddenly closed down his popular International Flying Saucer Bureau (IFSB). In the last issue of the bureau's magazine, *Space Review* (October), he included a puzzling statement. He said that he now knew the answer to the UFO mystery but could not publish it because of "orders from a higher source." He also urged "those engaged in saucer work to please be very cautious."

When further questioned by Gray Barker, who had been IFSB's chief investigator, Bender would only say that three men in black suits had visited him in September, told him what UFOs were, and threatened him with prison if he revealed this information. He told Barker that the strangers were "members of the United States government." So disturbing was the whole experience that Bender soon fell ill.

The exact nature of the men in black—or, as they would eventually be called in UFO circles, MIB—grew more unclear each time Bender reluctantly told his story. Soon some suspected that the MIB were not American intelligence agents but alien beings. Barker produced a frightening book about the event, *They Knew Too Much about Flying Saucers,* in 1956 and over the next few years wrote about the "Bender mystery" in a number of publications. In them, Bender's visitors were described as evil humans, aliens, or even demons.

In 1962 Bender wrote—and Barker published—*Flying Saucers and the Three Men.* It was a wild story that few readers accepted as true. In it Bender recalled being taken to the South Pole by monstrous aliens, who then followed his activities until 1960, when they returned to their home planet.

Men in Black Driving Cadillacs

Men-in-black stories appeared again in the 1960s. New York writer John A. Keel reported that UFO witnesses in New York, Ohio, West Virginia, and elsewhere had been accosted by MIB. Keel even claimed meetings of his own: "I kept rendezvous with black Cadillacs on Long Island and when I tried to pursue them, they would disappear impossibly on dead-end roads.... More than once I woke up in the middle of the night to find myself unable to move, with a dark apparition standing over me." According to Keel, the MIB were not government agents or even human beings but alien representatives. Often described as Oriental-looking, they behaved strangely, asking odd or even rude questions of those they visited. They usually traveled in large black cars.

Keel warned investigators: "Do not attempt to apprehend MIB yourself. Do not attack them physically. Approach them with great caution. They frequently employ hypnotic techniques." He also felt that the dangers of MIB visitation were so great for the weak-minded and the young that parents should "forbid their children from becoming involved [in UFO study]. Schoolteachers and other adults should not encourage teenagers to take an interest in the subject."

Meetings with MIB were not reported only by Keel's witnesses or just in the United States. In May 1975, two weeks after a dramatic UFO sighting from his plane (also recorded on radar screens at the Mexico City airport), a young pilot was on his way to a television interview when four black-suited men in a black limousine chased him down the freeway. After forcing him to the side of the road, they warned him not to discuss his sighting. A month later one of the strangers reappeared and threatened him again as he was on his way to a hotel to meet with J. Allen Hynek, the U.S. Air Force's top UFO adviser. That was the pilot's last encounter with the MIB, whom he remembered as tall and strangely white. He added: "I never saw them blink."

By the late 1980s, so many tales of MIB had been reported that the subject was included in the *Journal of American Folklore*. The author, Peter M. Rojcewicz, looked at the role of MIB in flying-saucer legends and related it to demon sightings reported in generations past. He also told of his own MIB experience. While doing research on UFOs in a library, he was approached by a dark-suited man who, speaking briefly in a slight accent about flying saucers, placed his hand on Rojcewicz's shoulder and said, "Go well in your purpose" and disappeared.

Sources:

Barker, Gray, *They Knew Too Much about Flying Saucers,* New York: University Books, 1956.

Bender, Albert K., *Flying Saucers and the Three Men,* Clarksburg, West Virginia: Saucerian Books, 1962.

Keel, John A., *UFOs: Operation Trojan Horse,* New York: G. P. Putnam's Sons, 1970.

UMMO

The Ummo affair was one of the strangest, most complex UFO hoaxes ever recorded. It began in 1965, when Fernando Sesma—a "contactee" who directed the Society of Space Visitors—reported receiving a phone call from a man who spoke in Spanish. The caller said he represented an "extraterrestrial order." While refusing to meet with Sesma, he promised that he would contact him again.

Soon afterward Sesma and other society members began receiving documents in the mail. They were supposedly written by residents of Ummo, a planet said to revolve around the star Iumma, 14.6 light–years from the sun. Each document bore an unusual symbol that looked like this:)+(. While astronomers stated that neither Ummo or Iumma existed, a group of believers soon sprang up in Europe and North and South America.

On February 6, 1966, several soldiers and two other witnesses saw a large circular object touch down briefly in a Madrid, Spain, suburb. One observer who caught a glimpse of the UFO's underside saw an odd symbol. It was the secret Ummo sign.

An advertisement in the May 20, 1967, issue of the Spanish newspaper *Informaciones* announced that on June 1 an Ummo craft would land outside Madrid and carry some faithful believers to the home planet. On that date, in the Madrid suburb of San Jose de Valderas, a flying object bearing the Ummo symbol was seen by a number of witnesses. Afterward, two sets of photographs of this UFO surfaced, but they were soon exposed as fakes. The craft displayed in the photos was actually a model made from plastic plates, about eight inches across. The symbol was scrawled on the bottom of the model with a marker pencil.

Later that same evening, in the Madrid suburb of Santa Monica, witnesses reported seeing a UFO approach the ground, then take off and disappear. The next morning signs of a landing were found at the site. So were metal cylinders that, when opened, contained the Ummo symbol. When studied in a scientific laboratory, though, these otherworldly objects proved to be made of earthly materials. One investigator thought that the UFO seen in the two sightings might be a radio-controlled model, for he noted that both cases took place near an airport and the Aerotechnical School, "both of which would have been convenient places to build, control, and hide a disk-shaped model."

Meanwhile, growing numbers of Spanish ufologists were getting Ummo documents. By the end of the decade, at least 600 pages on Ummo science, philosophy, politics, and civilization had been collected. The documents arrived through the regular mail, and most had a Madrid postmark. But others were sent from Australia, New Zealand, England, Argentina, France, Czechoslovakia, Yugoslavia, and the United States.

These documents suggested that the author or authors were highly educated and conversant in physical and biological science. But to most who read them, they were clearly not the reflections of an advanced extraterrestrial race. As Jacques Vallee, a ufologist trained in astrophysics and computer sciences, attested, "The Ummo technology is without major surprises, and it matches the kind of clever extrapolations one finds in any good science-fiction novel."

> One observer who caught a glimpse of the UFO's underside saw an odd symbol. It was the secret Ummo sign.

Ummites

The documents revealed that the Ummites, who look much like humans, arrived on Earth on April 24, 1950, when they landed in the French Alps. Since that time they have been observing Earth's affairs, though not interfering in them. They communicate with each other mentally—by telepathy—because at age 14 their vocal cords close up. Most of their documents were written in Spanish, but a few were in French.

The cult of Ummo believers is worldwide. Collections of Ummite documents have been published in Spanish and English, and letters from the Ummites continue to appear. According to Spanish ufologists Carlos Berche Cruz and Ignacio Cabria Garcia, Sesma and people within his organization (where the whole Ummo affair began) are quite likely perpetrators of a widespread hoax. Still, the charge can't be proven. They point out, though, that in the 1950s Sesma's writings about his space contacts included many concepts that suspiciously resembled those found in the Ummo documents.

A GOVERNMENT HOAX?

Some writers have suggested, though without evidence, that the Ummo affair is an experiment designed to explore human nature and society, run secretly by a government intelligence agency.

Sources:

Vallee, Jacques, *The Invisible College: What a Group of Scientists Has Discovered about UFO Influences on the Human Race,* New York: E. P. Dutton and Company, 1975.

Vallee, Jacques, *Revelations: Alien Contact and Human Deception,* New York: Ballantine Books, 1991.

Other Worlds

- HOLLOW EARTH
- FOURTH DIMENSION
- MOON ODDITIES
- VULCAN

Other Worlds

HOLLOW EARTH

The notion that the Earth has a hollow interior in which an underground civilization lives is an old one. Some would say that religious beliefs in hell and Hades are expressions of the concept. But the first American to try to prove the idea was the eccentric John Cleves Symmes (1779-1829). Symmes believed that the Earth was made up of a series of spheres, one inside the other, with 4,000-mile-wide holes at the North and South Poles. In spite of great ridicule, Symmes wrote, lectured, and worked hard to raise money for an expedition through the poles to the interior. There he planned to meet the inner-earth people and open "new sources of trade and commerce."

To the rest of the world, Symmes is remembered, if at all, as the inspiration for nineteenth-century American writer Edgar Allan Poe's early science-fiction tale of a hollow earth, *The Narrative of Arthur Gordon Pym* (1838). Yet Symmes was a pioneer of sorts, a man who paved the way for generations of freethinkers who imagined a new earthly geology and dreamed of another race secretly sharing the planet with us.

Tropical Splendor vs. Fiendish Race

By the late nineteenth century, there were many hollow-earth believers, especially after Symmes's son Americus published a collection of his father's lectures. While scientists dismissed the idea as absurd and physically impossible, that did not keep writers from detailing their own versions of the concept. Marshall B. Gardner, the author

In his 1974 book *Secret of the Ages,* Brinsley le Poer Trench reported that evil inner-earthers regularly kidnapped surface people and brainwashed them into becoming their agents.

Marshall B. Gardner's
version of the hollow earth.

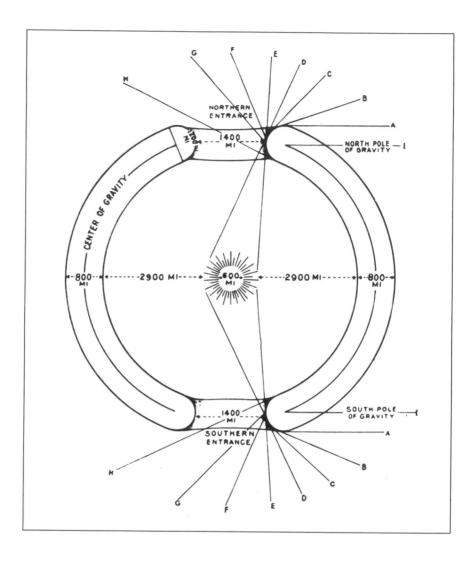

of the 1913 book *A Journey to the Earth's Interior,* for example, believed that there was a sun inside the Earth. Six hundred miles across, it gave the underworld a pleasant climate, allowing its inhabitants to live in tropical splendor! Guy Warren Ballard, who wrote *Unveiled Mysteries* (1934) under the name Godfre Ray King, told of the many out-of-body tours he had had beneath the Earth, led by an immortal "Master." There he found a beautiful world of scientific and spiritual marvels. And in time he even began meeting space people!

Nearly all nineteenth- and twentieth-century hollow-earth believers spoke of the inner world's inhabitants as members of an advanced, kindly race whose dealings with human beings could bring only good.

That is, until Richard Sharpe Shaver came along, describing his frightening experiences with a fiendish race beneath the earth in magazines like *Amazing Stories*. Called "deros," the evil creatures Shaver described (allegedly descendants of the people of Lemuria, whose continent was thought to have sunk into the Pacific Ocean thousands of years ago) had advanced technology that they used to torture kidnapped surface people. Deros also used their machines to cause accidents, madness, and other miseries in the world above.

The Nazis and the Hollow-Earth Theory

There is even a Nazi hollow-earth theory. According to Canadian Nazi sympathizer Ernst Zundel, who writes under the name Christof Friedrich, Nazi German dictator Adolf Hitler and his Last Battalion escaped to Argentina in a submarine as World War II ended. They then set up a base for advanced saucer-shaped aircraft inside the hole at the South Pole. To Zundel the Nazis were "outer earth representatives of the 'inner earth'"—which is one fantastic way of justifying Nazi claims of racial superiority.

Brinsley le Poer Trench wrote about the terror of this hidden world; in his 1974 book *Secret of the Ages,* Trench reported that evil inner-earthers regularly kidnapped surface people and brainwashed them into becoming their agents. Now, he said, the "ground work has ... been prepared for a takeover of this planet by those who live inside it."

But by far the most popular book on the subject was Walter Siegmeister's *The Hollow Earth* (1964), which he wrote under the pseudonym Raymond Bernard. It contributed little that was new to the inner-earth legend, quoting mostly from nineteenth-century texts on the matter. Still, it did introduce the author's belief that there existed a conspiracy to hide the truth about the hollow earth, flying saucers, and pole holes. The book sold well, going through numerous printings, and exposed many readers to the idea of a hollow earth for the first time.

REEL LIFE

At the Earth's Core, 1976.

A Victorian scientist invents a giant burrowing machine, which he and his crew use to dig deeply into the earth. To their surprise, they discover a lost world of subhuman creatures and prehistoric monsters.

Journey to the Center of the Earth, 1959.

A scientist and student undergo a hazardous journey to find the center of the earth. Along the way they find the lost city of Atlantis. Based upon the novel by Jules Verne.

Through the years, numerous expeditions have been planned to the North and South Poles to attempt to enter the inner earth. Many UFO enthusiasts, occultists, religious movement founders, and even quite a few contemporary pro-Nazis have advocated the hollow-earth theory as support for their views.

Sources:

Bernard, Raymond (pseudonym for Walter Siegmeister), *The Hollow Earth: The Greatest Geographical Discovery in History,* New York: Fieldcrest Publishing, 1964.

Friedrich, Christof (pseudonym for Ernst Zundel), *UFOs—Nazi Secret Weapons?,* Toronto: Samisdat, 1976.

Michell, John, *Eccentric Lives and Peculiar Notions,* San Diego, California: Harcourt, Brace, Jovanovich, 1984.

WHAT IS DIMENSION?

Dimension signifies a position in space. Space, as science has generally defined it, is three-dimensional; it is bound by width, height, and length. When a person looks at a picture of an apple, he or she is seeing only two dimensions, because the picture is flat. Its depth cannot be seen or felt. But when touching an actual apple, the observer can feel the three elements of its width, height, and length.

We can conceive of the three dimensions, but what about a fourth—or even a fifth—dimension? These can only be imagined. Other dimensions either do not exist, as many scientists believe, or most human beings are not equipped to experience them.

FOURTH DIMENSION

In February 1945, N. Meade Layne, a San Diego man interested in the supernatural, founded Borderland Sciences Research Associates (later the Borderland Sciences Research Foundation). The organization was devoted to Layne's belief in "ethereans," beings like us who live in another dimension. Layne explains that "just as there is a spectrum of sound and color (ending in sounds we cannot hear and colors we cannot see)," there is also a spectrum of matter—of places and people—of which we ordinarily have no awareness. This matter is, in fact, too dense to be touched; that is, according to Layne, until ethereans lower their atomic vibratory rates—and we can see or feel them.

Ethereans and UFOs

Layne believed that it is ethereans who pilot flying saucers. More advanced than us, the ethereans nonetheless live in a world much like our own—almost a mirror reflection. The stars and planets of our universe have their etheric doubles. As the theory goes, ethereans are really just fourth-dimension human beings.

Other Worlds

While Layne did not invent the idea of an etheric world, he was the first to link it with flying saucers. By working with a psychic medium who supposedly received communications from ethereans, Layne claimed to learn the secrets of that parallel world's science and philosophy.

Layne's "etherean ships" appealed only to those interested in the most fantastic ideas surrounding **UFOs.** Still, the thought of unidentified flying objects coming from another dimension did gather interest. Next to the "beings from outer space" idea, the fourth dimension (4-D) theory became ufologists' favorite UFO explanation; though seldom credited, Layne's ideas had a marked impact on how ufologists perceived UFOs.

Other writers offered their own versions of the connection between UFOs and a dimension outside our own. In his 1970 book *UFOs: Operation Trojan Horse,* for instance, John A. Keel declared that shape-changing "ultraterrestrials" from other levels of reality "can make us see what they want us to see and remember only what they want us to remember." Similar to Layne's ethereans, these ultraterrestrials could appear in our world by "manipulations of electromagnetic energy," entering "a solid state" that would even allow them to "leave impressions on the ground." But unlike Layne's harmless ethereans, Keel's ultraterrestrials were up to no good. Secret rulers of the earth for many centuries, they still have us under their control: "We are biochemical robots helplessly controlled by forces that can scramble our brains, destroy our memories and use us in any way they see fit," wrote Keel. "They have been doing it to us forever."

The idea of a fourth dimension intrigued writers because it suggested an order or control system beyond our own. Maybe strange events that were unexplainable in our world did originate from another place, where they made perfect sense. Some writers combined two terms, *paranormal* (outside the normal) and *physical,* to create a new word, "paraphysical." This concept encompasses occurrences once called "etheric"—events that have both natural (like leaving tracks) and unnatural (like disappearing instantly) features.

Sources:

Keel, John A., *Strange Creatures from Time and Space,* Greenwich, Connecticut: Fawcett, 1970.

Keel, John A., *UFOs: Operation Trojan Horse,* New York: G. P. Putnam's Sons, 1970.

Layne, N. Meade, *The Ether Ship and Its Solution,* Vista, California: Borderland Sciences Research Associates, 1950.

> Next to the "beings from outer space" idea, the fourth dimension (4-D) theory became ufologists' favorite UFO explanation.

REEL LIFE

The 4D Man, 1959.

A physicist's special project gets out of control, leaving him able to pass through matter and see around corners. He also finds that his touch brings instant death.

MOON ODDITIES

"Transient lunar phenomena," or TLP, are short-lived oddities that appear on the moon's surface, observed by professional or amateur astronomers through telescopes or, more rarely, with the naked eye. TLP include dark spots, lights, and moving objects. Dark spots are sometimes explained as optical effects, shadows of moon features like mountains or valleys, for example. Some of the lights may be no more than reflections from the sun. And objects observed moving across the moon's surface may actually be earthbound things, like birds or seeds, that enter a witness's field of vision and mistakenly *appear* to be located in outer space. Meteors passing between the earth and the moon may also account for some TLP sightings.

Between 1540 and 1970, over 900 reports of TLP were recorded. These include sightings of reddish patches, clouds, and flashes on the moon. And while astronomers have been able to supply reasons for many of these lunar oddities, even they admit not all TLP can be easily explained.

Extraterrestrials on the Moon

Most scientists feel that sightings of TLP do not mean that intelligent activity—carried out by alien beings—is taking place on the moon. Still, quite a few books have been written about the subject, and their popularity has demonstrated that the public is receptive to the idea.

Perhaps the most famous person to believe in intelligent life on the moon was the great British astronomer Sir William Herschel (1738-1822). Though best remembered as the discoverer of Uranus, Herschel noted in a paper read to the Royal Society in 1780, "the great probability, not to say almost absolute certainty, of [the moon] being inhabited." His observations of both TLP and the moon's ordinary surface features led him to believe that he had seen cities, towns, roads, canals, forests, and even circuses there! Herschel's reputation perhaps survived these outlandish notions because he never went so far as to publish them; he recorded them only in his private journal, which has been discovered just recently.

In stories published between August 25 and 31, 1835, the *New York Sun* reported that Sir John Herschel, Sir William's son, observed two species of "bat-men" on the moon. The simpler creatures averaged "four feet in height" and were "covered, except on the face, with short and glossy copper-colored hair" and had "wings composed of a thin

The *New York Sun* reported that Sir John Herschel, Sir William's son, observed two species of "bat-men" on the moon.

Sir William Herschel, astronomer, discovering Uranus, assisted by his sister.

membrane, without hair, lying snugly upon their backs, from the top of the shoulders to the calves of the legs." Their faces resembled those of orangutans. The higher species of lunar bat-men were "of infinitely greater personal beauty ... scarcely less lovely than ... angels."

Later, *Sun* reporter Richard Adams Locke admitted he had made up the story. His trick, referred to as the Moon Hoax, had fooled thousands of readers, even those who were well educated and should have

Herschell recorded observing cities, forests, and even circuses on the moon.

known better. Writer Edgar Allan Poe noted with astonishment that "not one person in ten discredited it." Was Locke a true hoaxer, playing his trick to gain attention? Michael J. Crowe, a historian of science, did not believe this perception of Locke, but thought Locke only intended to poke fun at the popular belief that life existed in other worlds. He thought that his readers would laugh at his account. Imagine his surprise when they accepted it as true!

TLP, Extraterrestrials, and UFOs

Charles Fort, one of the first to research and catalog reports of strange events, believed that intelligent beings from other worlds did exist. From scientific journals he collected many accounts of strange lights and oddities on or near the moon, and sometimes he would connect these with unusual events that took place on the earth, arguing that beings from outer space were responsible for both. Fort believed that otherworldly beings were always watching us. After his death in 1932, Fort's followers continued to collect reports supporting the idea of "space visitors," and science-fiction magazines like *Amazing Stories* also ran articles on the subject.

In July 1951 *Fate,* a magazine about incredible "true mysteries," published the account of George Adamski of Palomar Gardens, California, describing his success photographing spaceships through his six-inch telescope. Adamski wrote:

> I have taken all my pictures at night by the light of the moon because often I had noticed that a good number of the ships I saw moving through space appeared headed for the moon. Some of them seemed to land on the moon, close to the rim; while others passed over the rim and disappeared behind it....
> I figure it is logical to believe that space ships might be using our moon for a base in their interplanetary travels.

On November 20, 1952, Adamski reported that he actually met a spaceship pilot, from Venus, in the desert of southern California. He would later claim many more meetings with beings from Venus, Mars, and Saturn. One such contact occurred in August 1954, when Adamski insisted that a Venusian scout craft flew him around the moon, where he saw extraterrestrial cities and spaceship hangars as well as forests, lakes, and rivers (also see entry: **Space Brothers**).

Adamski published a book about this moon adventure, *Inside the Space Ships,* in 1955. Most people, including a great many ufologists, concluded that Adamski's otherworldly experiences existed only in his head. Still, some did believe the author and tried to provide "proof" for his claims. Fred Steckling, head of the George Adamski Foundation, wrote the book *We Discovered Alien Bases on the Moon* (1981), which lent support to Adamski's ideas. In it Steckling insists that the U.S. government is involved in a huge cover-up—hiding the fact that both space people and an atmosphere exist on the moon.

Like Adamski, others claimed contact with space visitors during the 1950s and 1960s and told of bases and beings on the moon. One of them, Howard Menger, even claimed to have taken a photograph there. His 1959 book *From Outer Space to You* contains a picture that claims to show a "spacecraft landing near dome-shaped building." Ozark farmer Buck Nelson also claimed that he landed on the moon one April day in 1955, in the company of his dog Teddy, a Venus resident named Little Bucky, and Little Bucky's dog Big Bo. Buck observed children playing with several dogs while on the moon. Then the group took off for Venus!

Even those who studied UFOs seriously—and rejected the outlandish stories of "contactees"—were intrigued by the idea of extraterrestrials on the moon. Donald E. Keyhoe, a retired marine corps major, aircraft writer, and ufologist, for example, reacted with great excitement when two amateur astronomers observed, in 1953, what they thought was a natural bridge that suddenly appeared near the moon's Mare Crisium crater. Keyhoe wrote that the bridge's sudden appearance ruled out a natural explanation and that "evidence of some intelligent race on the moon seemed undeniable." (Astronomers have since decided that the "bridge" is an optical illusion.)

ATLANTIS

Atlantis is a large island in Greek legend located in the Atlantic Ocean west of Gibraltar. The Greek philosopher Plato admired the Atlanteans' educated and cultured way of life, which was snuffed out when an earthquake caused the island to be swallowed up by the sea. Some still believe in the legend today, and societies for the rediscovery of Atlantis actively look for remains of the civilization.

Another writer of the period, M. K. Jessup, had an even more fantastic idea concerning beings on the moon. In his book *The Expanding Case for the UFO* (1957) he wrote that tens of thousands of years ago pygmy races—much older than other human races—developed antigravity spaceships and escaped to the moon just as natural disasters wiped out other advanced civilizations, like the mythical Atlantis. The pygmies continue to observe us from their present home, thus explaining both UFOs and the "little men" often seen with them (also see entries: **Ancient Astronauts** and **Philadelphia Experiment**).

Clear photographs of the moon's surface became available to the public following the U.S. lunar explorations that took place between 1969 and 1972. Some who saw the pictures believed that they showed evidence of extraterrestrial activity; George H. Leonard, who wrote *Somebody Else Is on the Moon* (1976), and Don Wilson, author of *Our Mysterious Spaceship Moon* (1975), were the best known of these individuals. They accused NASA (the National Aeronautics and Space Administration) of trying to hide this fantastic discovery. Furthermore, Wilson believed not only that spaceships were on the moon, but that the moon itself *was* a spaceship—a huge hollow harbor built thousands of years ago by an alien race.

Responding to the growing number of writings about extraterrestrial activity on the moon, Francis Graham of the Pennsylvania Selenological Society (a branch of astronomy that deals with the moon) took a look at all the claims. While open-minded on the subject of UFOs, he concluded that "there is not a single piece of unambiguous [clear] evidence for the existence of alien bases on the moon." The astronomer felt that people who claimed to see evidence in NASA lunar pho-

Astronauts walk on the moon.

tographs reached false conclusions because they were unfamiliar with the moon's natural features and geological oddities, and the photographs they used were often unclear.

The Apollo Aliens

Shortly after the first moon landing, on July 16, 1969, the *National Bulletin,* a supermarket tabloid, published a tall tale that captured the public's imagination. According to the story, the Apollo 11 astronauts saw spaceships when they arrived on the moon, and NASA managed to censor their radio report so that the news media, and therefore the rest of the world, would not learn of the shocking discovery. Yet somehow someone had slipped a tape of the astronauts' message—reporting the appearance of two UFOs along a crater rim—to the *Bulletin.*

Four years later, Stuart Nixon of the National Investigations Committee on Aerial Phenomena looked into the story and found, not to his surprise, that it contained no shred of truth. The printed record of the supposed conversation between the astronauts—Neil Armstrong, Edwin "Buzz" Aldrin, and Michael Collins—and Mission Control contained so many errors in fact and terminology that it had to be false. And no such conversation could have been censored at the time it was sent, for even a short break in communication would have been noticed immediately. The *Bulletin* could not produce the original tape of the message, nor could it even produce the reporter, Sam Pepper, who wrote the account.

Though it appeared that the story was simply made up in the *Bulletin* office, it would be retold again and again—in one form or another—and become a part of Space Age folklore. Maurice Chatelain wrote about it in his 1978 book *Our Ancestors Came from Outer Space;* he claimed to get his information from sources within NASA. Charles Berlitz and William L. Moore discussed it in 1980 in their book *The Roswell Incident.* According to their sources, which they did not make altogether clear, "NASA was forced to change the originally intended landing site for the Eagle lander module because it was discovered that the first site was 'crawling'—presumably with somebody else's space hardware." Berlitz and Moore were later sued by astronaut Aldrin for writing a false story about him.

Sources:

Adamski, George, *Inside the Space Ships,* New York: Abelard-Schuman, 1955.
Crowe, Michael J., *The Extraterrestrial Life Debate 1750-1900: The Idea of a Plurality of Worlds from Kant to Lowell,* New York: Cambridge University Press, 1986.
Graham, Francis G., *There Are No Alien Bases on the Moon,* Burbank, California: William L. Moore Publications and Research, 1984.

VULCAN

In 1846 Urbain Leverrier of the Paris Observatory was one of two astronomers to suggest that an eighth planet existed in the outer reaches of our solar system. Disturbances in the orbit of Uranus led Leverrier to believe that another large heavenly body was causing the irregularities; from these he was able to calculate almost exactly where the new planet could be found. Others duplicated his calculations and discovered that an eighth planet did, indeed, exist.

Other Worlds

Leverrier, who had a very high opinion of himself, wanted the new planet named after him, but it was soon called Neptune. He also fought to exclude British astronomer John Adams from any recognition for the discovery, even though Adams had made similar calculations concerning an eighth planet. Many, in fact, thought that the two scientists should have been named Neptune's co-discoverers.

Perhaps Leverrier's hunger for honor and fame explains his odd behavior a few years later. The French astronomer had begun to focus his attention on the opposite end of the solar system, toward Mercury, which, like Uranus, had its own orbit irregularities. Though the theory of relativity would later provide an explanation for these, in Leverrier's time the only cause he could imagine was an intra-Mercurial planet—in other words, a world in orbit between Mercury and the sun.

Urbain Leverrier, French astronomer.

The New Planet

On December 22, 1859, Leverrier received a letter from a country doctor and astronomy hobbyist named Lescarbault. The man made an extraordinary claim: that on March 26, 1859, he had seen a round black spot—a planet—move across the upper part of the sun's face. Leverrier immediately went to the village of Orgeres, where Lescarbault lived. Without identifying himself, the astronomer badgered the physician with questions and even made fun of him, but Lescarbault stood by every detail of his story. Finally, Leverrier revealed who he was, warmly congratulated the physician, and on his return to Paris saw to it that Lescarbault would be decorated with the Legion of Honor.

Within days the new discovery had the world of astronomy buzzing. Leverrier, perhaps more careful after the unpleasant competition surrounding the detection of Neptune, suggested that the planet be named Vulcan. By January, excited discussions about the discovery were appearing in astronomy journals. Leverrier calculated the new planet's size (about 1/17th that of Mercury, he thought) and guessed

The solar
system.

that it crossed the sun's face in early April and early October. He also cited 20 earlier sightings, which he now felt sure identified Vulcan.

From the beginning, however, there were doubters. One was a Brazilian astronomer who reported observing the sun's face at the same time as Lescarbault. The Brazilian had seen nothing out of the

Other Worlds

ordinary, and his telescope was much more powerful than the good doctor's! Over the next few decades, astronomers watched for Vulcan during the periods Leverrier predicted it would appear. Results were disappointing, with most sightings really observations of sunspots. By the end of the century, there was scarcely a Vulcan believer left. In 1899 Asaph Hall, the discoverer of the moons of Mars, remarked that the planet was no longer a part of "rational astronomy."

Still, some of the sightings that pointed to the existence of Vulcan were, and remain, puzzling. Perhaps the mysterious objects that astronomers observed near the sun were really much closer to them than they thought (UFOs?) and that was why observers at other locations could not duplicate their findings. Such would seem the case with two U.S. astronomers (one in Wyoming and one in Colorado) who observed two shining objects some distance from the sun during a total solar eclipse on July 29, 1878. No one else reported the phenomenon, and their reports caused heated discussions among other astronomers, who accused the two of making the simplest of errors: mistaking two well-known stars for two unknown objects. Nevertheless, the two observers rejected the accusations. "I have never made a more valid observation," one of them, Lewis Swift, wrote in *Nature,* "nor one more free from doubt."

SUNSPOTS

Sunspots are dark, usually irregularly shaped patches that appear in groups on the sun's surface and are actually magnetic storms. Periods of great sunspot activity usually occur in cycles of 11 years. During these times, various corresponding disturbances take place on earth, such as magnetic storms, faulty radio reception, and malfunctioning magnetic compasses.

Sources:

Corliss, William R., ed., *Mysterious Universe: A Handbook of Astronomical Anomalies,* Glen Arm, Maryland: The Sourcebook Project, 1979.
Fort, Charles, *The Books of Charles Fort,* New York: Henry Holt and Company, 1941.
Grossinger, Richard, *The Night Sky,* Los Angeles: Jeremy P. Tarcher, 1981.

Government Cover-ups

- HANGAR 18
- PHILADELPHIA EXPERIMENT
- AREA 51

Government Cover-ups

HANGAR 18

I n the 1960s Arizona senator and U.S. Air Force Reserve brigadier general Barry Goldwater asked a friend, General Curtis LeMay, for a favor. Senator Goldwater wanted to see a room at Wright-Patterson Air Force Base in Dayton, Ohio, where UFO wreckage and deceased UFO pilots were rumored to be secretly stored. As the senator recalled years later in a *New Yorker* article, General LeMay "just gave me holy hell. He said, 'Not only can't you get into it but don't you ever mention it to me again.'"

In the summer of 1947, not long after pilot Kenneth Arnold sighted flying saucers over Mount Rainier, Washington (also see entry: **Unidentified Flying Objects**), the world press reported that army air force workers had recovered the remains of a "flying disc" that had crashed in remote Lincoln County, New Mexico. Within hours a "correction" went out over the wires, with army officials assuring reporters that the story had been a mistake, that the wreckage of a weather balloon had been misidentified as something extraordinary.

The explanation—now known to be false—was widely accepted at the time, and the story died a quick death. The event was rarely mentioned again, that is, until ufologists (researchers of unidentified flying object sightings) began a reinvestigation of the case in the late 1970s. By 1992 four books had been written about "the Roswell incident" (so named because the recovery operation took place at the army air force base in Roswell, New Mexico), and the investigation continues to this day.

Army Finds Air Saucer On Ranch in New Mexico

Disk Goes To High Officers

Picked Up
Last Week

ROSWELL, N.M. — (AP) — The Army Air Force here today announced a flying disk had been found on a ranch near Roswell and is in Army possession.

Lt. Warren Haught, public information officer of the Roswell Army Air Field, announced the find had been made sometime last week, and had both turned over to the airfield through operations of the sheriff's office.

It was rumored at the Roswell Army Air Field and subsequently named by Maj. Jesse A. Marcel of the 509th bomb group intelligence office at Roswell. The higher headquarters.

The Army gave no other details.

'Flying disc' turns up as just hot air

Fort Worth, Tex., July 9 (AP).—An examination by the Army revealed last night a mysterious object found on a lonely New Mexico ranch was a harmless high-altitude weather balloon — not a grounded flying disc.

Army Knocks Down Disk—

IT'S A WEATHER BALLOON

Device Is Only A Wind Target

Object Found in N. Mexico
Identified at Fort Worth

More on Page 1

FORT WORTH — A flying disk reported to the Army Air Forces to have been found near Roswell, N.M., was today stripped of its glamor tonight by a Fort Worth Army airfield weather officer. He identified the object as a weather balloon.

Warrant Officer Irving Newton, a weather forecaster at the base weather station, said the device was a balloon and a target used to determine the direction and velocity of winds at high altitudes.

80 Stations Use Same Type Balloon

Newton said there were some 80 weather stations in the United States using this same type balloon and that there were some from any one of them.

The balloon was shipped to the Fort Worth head quarters here from Roswell and officers were shocked it was to have been forwarded to Wright Field near Dayton.

Warren Haught, public information officer at Roswell, announced earlier in the day that the 509th bomb group intelligence office had taken possession of a mysterious object found near the Army field. The Army base at Fort Worth identified the object as a weather balloon.

According to some informants, searchers found the bodies of four gray-skinned humanoids at a location two miles from the main crash site. They reported that officials swore all who knew of the event, military or otherwise, to secrecy. Decades later investigators would find witnesses and participants who still would not discuss what they knew. Regardless, by the early 1990s, ufologists like William Moore, Stanton Friedman,

Government Cover-ups

Kevin Randle, and Don Schmitt had collected the testimonies of several hundred people—from local ranchers to air force generals—and from these reconstructed what they believed had occurred. The Roswell incident has become perhaps the best-documented case in UFO history.

Unearthly Rumors

Before this wide-scale investigation, ufologists had heard rumors about "little men in pickle jars"—preserved remains of alien crash victims—but could not find any real proof. They were concerned that such rumors were merely hoaxes, like the one con artists Silas Newton and Leo A. GeBauer used to scam unlucky author Frank Scully. Claiming to know of spaceship crashes in the Southwest in order to sell phony oil-detection devices based on "extraterrestrial" technology, the two related their bogus experiences to Scully, who recorded them in the popular book *Behind the Flying Saucers* (1950). Later exposed for the shady characters they were, Scully's informants made him look like a trusting fool.

So that hoax, along with a lack of evidence of other claimed crashes and recoveries, made most ufologists doubt that UFO wrecks and corpses existed. As Ed J. Sullivan stated in the September 1952 issue of *Civilian Saucer Investigation Quarterly Bulletin,* crash stories "are damned for the simple reason, that after years of circulation, not one soul has come forward with a single concrete fact to support the assertions. If there were one single iota of fact, certainly someone, somewhere, would be willing to bring it into the open."

Still, rumors continued, many focusing on the crash landings of one or more spacecraft in the Southwest. Military workers reportedly transferred the wreckage and bodies to Wright-Patterson Air Force Base in Dayton, Ohio, where the air force had set up its first UFO investigation unit. In some cases, individuals said that they had actually seen the evidence, either at the recovery site or—accidentally—in a secret room at Wright-Patterson. At some point this room came to be known as "Hangar 18" or "Blue Room." A 1980 science-fiction film, *Hangar 18,* dramatized the story. Of course, the air force has long denied that any such room exists.

In the 1970s well-known ufologist Leonard H. Stringfield decided that, unlike his doubting colleagues, he would actively seek out crash testimony, including Hangar 18 stories. He published his findings in a series of reports, with the most interesting accounts coming from eyewitnesses, some of which are excerpted here.

To many UFO buffs, this widely circulated photograph represents a picture of an extraterrestrial in a secret U.S. government vault. In fact, it is a wax dummy displayed in Canada in the early 1980s.

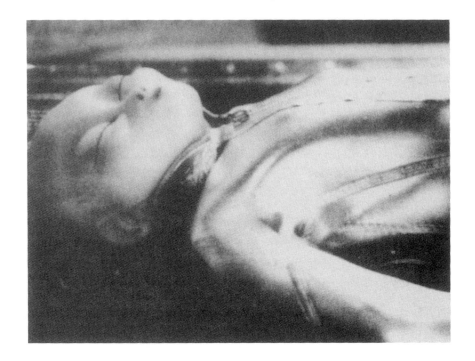

Humanoid Corpses Seen in 1953

A former army pilot said he had been inside a Wright-Patterson hangar one day in 1953 when a DC-7 arrived with five crates. The informant saw three of them opened; inside each lay the body of a small humanoid on fabric stretched over a bed of dry ice. Four feet tall, the beings had large, hairless heads and small mouths; the bodies were thin and looked brown under the hangar lights. They were dressed in tight-fitting uniforms, and one, with two bumps on the chest, appeared to be female. The pilot said crew members from the DC-7 later told him that a flying saucer had crashed in the Arizona desert and that one of the humanoids was still alive when a recovery team arrived, which tried without success to save it.

Heavily Guarded Bodies Seen in 1966

A retired army intelligence officer stationed at Wright-Patterson in 1966 reportedly saw nine alien bodies in a heavily guarded section of the base. They were four feet tall and gray-skinned. He was told that a total of 30 such bodies were stored there, along with the wreckage of spacecraft. The officer also revealed that "since 1948 secret information concerning UFO activity involving the U.S. military has been contained in a computer center at Wright-Patterson AFB," with backup files stored at other military locations.

Another source, a retired air force pilot, told ufologist Stringfield that in 1952, while attending a "high-level secret meeting" at Wright-Patterson, he saw an alien body deep-frozen in an underground chamber. Four feet tall and hairless, it had a big head and long arms. He said he learned that some of the Wright-Patterson UFO material was eventually taken to the air force's underground complex at Colorado Springs, Colorado.

The most remarkable of Stringfield's witnesses was a medical doctor associated with a major hospital. The physician claimed that he had performed an autopsy on an alien body while serving in the military in the early 1950s. In a 1979 statement he prepared for Stringfield from his medical files he stated:

> The specimen observed was four feet three inches in length.... The head was pear-shaped in appearance and oversized by human standards for the body.... The ends of the eyes furthest from the nasal cavity slanted upward.... The eyes were recessed into the head. There seemed to be no visible eyelids, only what seemed like a fold. The nose consisted of a small fold-like protrusion.... The mouth seemed to be a wrinkle-like fold. There were no human type lips as such—just a slit that opened into an oral cavity about two inches deep.... The tongue seemed to be atrophied [wasted away] into almost a membrane. No teeth were observed.... The outer "ear lobes" didn't exist.... The head contained no hair follicles. The skin seemed grayish in color.

Perhaps what is most striking about these accounts is that they describe beings that look very much like those reported by UFO abductees—people who have claimed (or reveal under hypnosis) to have been kidnapped by aliens. Humanoids of this type are rare in early UFO accounts, which report spaceship occupants resembling humans or human dwarfs (also see entries: **Flying Humanoids** and **Hairy Dwarfs**), or in rare cases, monsters.

Stringfield had other medical sources who gave similar descriptions of alien corpses. But because the ufologist—as respected in the field as he was—insisted on keeping the names of his sources secret, the stunning testimonies he collected could be viewed only as stories; they could not be independently verified, a vital step if they were to become actual evidence.

More Than Rumors?

In other cases, however, names have been revealed, and these people tell tales similar to those related above.

Norma Gardner worked at Wright-Patterson for a number of years. Her high-level security clearance allowed her to view secret materials, including—as she once told Charles Wilhelm, a young friend interested in UFOs—items recovered from crashed spacecraft. She was responsible for cataloging, photographing, and tagging them. Once, she said, she saw two humanoid bodies being carted from one room to another. They were generally human in appearance, though they did have large heads and slanted eyes. She told Wilhelm that she was passing on the information only because she was dying of cancer and "Uncle Sam can't do anything to me once I'm in my grave."

One evening in July 1952, Pan American pilot William Nash and copilot William Fortenberry sighted a UFO while flying over Virginia. The next morning, as they waited for investigators from the air force's Project Blue Book to interview them, they agreed to ask if the rumors about crashed spacecraft at Wright-Patterson were true. Fortenberry got a yes answer from one of his interviewers. But when Nash asked a group of investigators whose commanding officer was present, he got a resounding "NO!" from the ranking official. It seemed to Nash that the response was meant less for him than to "shut up" the investigators who had already "opened their mouths to answer the question."

There have been other people who—while not claiming to have personally seen wreckage or bodies at Wright-Patterson—did acknowledge that such things existed. One was physicist Robert Sarbacher, who, in the late 1940s and early 1950s, was a consultant for the Defense Department's Research and Development Board (RDB). On September 15, 1950, during a meeting in his office with a group of Canadian government scientists and engineers, Sarbacher was asked if there was any truth to the constant rumors that the U.S. government was holding UFO remains. He answered yes, adding, "We have not been able to duplicate their performance.... All we know is, we didn't make them, and it's pretty certain they didn't originate on earth." The subject, he said, "is classified two points higher even than the H-bomb." He would say little more, except that a top-secret project had been formed to study the materials.

Ufologist Arthur Bray found evidence of this remarkable conversation three decades later in the personal papers of Wilbert B. Smith, a deceased Canadian radio engineer and UFO buff who had been at that September meeting with Sarbacher and had asked the key questions. As one would imagine, the discovery of these papers sent several investigators in search of Sarbacher, to see if what Smith had recorded was true. The physicist was located in Florida and admitted

that he had knowledge of spaceship crash recoveries from the official reports that had crossed his RDB desk. He recalled that "certain materials reported to have come from flying saucer crashes were extremely light and very tough," which is how witnesses described the material reportedly recovered in the Roswell incident. Sarbacher added that "there were reports that instruments or people operating these machines were also of very light weight, sufficient to withstand the tremendous deceleration and acceleration associated with their machinery." Even more remarkable, the doctor related that once he had been invited to a high-level meeting at Wright-Patterson, where air force investigators planned to discuss what they had learned from their work on wreckage and bodies stored there.

Perhaps more important still was the testimony of air force officer Arthur Exon. Exon stated in July 1947, when he was a lieutenant colonel at Wright Field (which became Wright-Patterson AFB), that the remains of a flying saucer and its occupants recovered in New Mexico were flown in from Eighth Army Headquarters to undergo study at the base's laboratories. Like other witnesses, he told investigators Randle and Schmitt that some of the material from the craft was "very thin but awfully strong and couldn't be dented with heavy hammers." He reported that all present at the time agreed that "the pieces were from space."

Science fiction or reality? A huge, hi-tech UFO project hidden in Middle America, in this scene from Steven Spielberg's *Close Encounters of the Third Kind.*

He noted that the bodies "were in fairly good condition." A top-secret committee took over the investigation of this and other UFO cases.

In 1964 Exon, by now a general, became base commander at Wright-Patterson. Even so, the area where secret UFO studies took place was off-limits to him, so he knew very little about them. He related that from time to time a "team of uniformed officers would arrive on a commercial [not military] flight" to talk with UFO project workers at the base before going out to investigate an important case. Then groups of them would board military aircraft Exon would make available to them and be gone for several days before returning. Exon was never told of their whereabouts. "We were never informed about any reports. They all went to Washington," he recalled.

Clearly, stories of this sort—and there are many more of them—seem to point to the existence of space visitors and of top secret government projects. Although the accounts do not prove anything by themselves, when gathered together they paint a convincing picture. Investigations into spaceship crashes, the Roswell incident, and Hangar 18 continue.

Sources:

Berlitz, Charles, and William M. Moore, *The Roswell Incident,* New York: Grosset and Dunlap, 1980.
Bernstein, Burton, "Profiles: AuH20," *New Yorker,* April 25, 1988, pp. 43-73.
Friedman, Stanton T., and Don Berliner, *Crash at Corona,* New York: Paragon Books, 1992.
Randle, Kevin D., and Donald R. Schmitt, *UFO Crash at Roswell,* New York: Avon Books, 1991.
Stringfield, Leonard H., *Situation Red, the UFO Siege!,* Garden City, New York: Doubleday and Company, 1977.

PHILADELPHIA EXPERIMENT

In October 1955 Morris Ketchum Jessup, author of a recently published book entitled *The Case for the UFO,* received a strange letter from a man named Carlos Miguel Allende. The poorly written, rambling letter told of a World War II experiment with physicist Albert Einstein's Unified Field Theory that made a ship—a destroyer and all of its crew members—completely invisible while at sea. This allegedly occurred in October 1943.

According to Allende, the experimental ship had also been transported from a dock in Philadelphia, Pennsylvania, to one in the Norfolk/Newport News/Portsmouth area of Virginia and back in a matter of minutes. Apparently, the temporary vanishing had caused most of the crew to go insane; at one point, while still invisible, they ransacked a tavern near the navy dock. "The expieriment [sic] Was a Complete Success," Allende wrote. "The Men Were Complete Failures."

Jessup paid little attention to this odd letter or to a second Allende sent in January 1956; that is, until he was invited to the Office of Naval Research (ONR) in Washington, D. C., where he found that ONR officers had received a copy of Jessup's book from an unknown sender. Puzzling comments about UFO intelligence and technology had been written in the margins of the book by three individuals in three different-colored inks. Jessup told the officers that the writings reminded him of Allende's, especially references to the Philadelphia experiment. For some reason, the officers became convinced that Allende's letters and the book notes were important and held possible "clues to the nature of gravity." They made a small number of copies of Jessup's book, this time including the notes and letters.

Suicide or Murder?

Meanwhile, personal difficulties unrelated to the Allende affair plagued Jessup. The three sequels he wrote to *The Case for the UFO* were poorly received. And he had money and marriage problems and was badly injured in a car accident. When he visited writer Ivan T. Sanderson in October 1958, he was deeply depressed. In April 1959 Jessup died of carbon-monoxide poisoning in his car in Dade County Park, Florida, after writing what appeared to be a suicide note to a friend. Authorities ruled that Jessup took his own life.

Few UFO buffs knew of the Jessup/Allende case until 1962, when Borderland Sciences Research Associates made the reprinted material widely available. Vincent Gaddis devoted a chapter in his 1965 book *Invisible Horizons* to the episode, and other writers soon did the same. In 1979 the case inspired a book of its own, William L. Moore's *The Philadelphia Experiment* (on which the 1984 science-fiction film of the same name was based). Anna Lykins Genzlinger provided more fuel for the growing modern legend with her book *The Jessup Dimension* (1981). She even speculated that government agents had murdered Jessup because he knew too much about the Philadelphia experiment.

It didn't seem to matter that no evidence could be found to support Allende's story. In fact, the Philadelphia experiment grew more fantastic with each retelling. Eventually the unlucky crew members were

Apparently, the temporary vanishing had caused most of the crew to go insane; at one point, while still invisible, they ransacked a tavern near the navy dock.

Morris K. Jessup

(1900-1959)

Morris Ketchum Jessup was born March 2, 1900, in Rockville, Indiana. He attended the University of Michigan in Ann Arbor as both undergraduate and graduate student in astronomy. There he participated in some groundbreaking astronomy projects.

From the 1930s through the 1950s Jessup spent much time in Mexico and Central and South America, where he observed archaeological sites with great interest. By the 1950s he was living in Washington, D.C., running an export business and writing about modern and ancient mysteries.

Jessup's first book, *The Case for the UFO* (1955), sought to link flying saucers with other strange or unexplained events and claims, including disappearances and falls from the sky. Two later works, *UFO and the Bible* (1956) and *The Expanding Case for the UFO* (1957), speculated about early human history, anticipating the themes of the ancient astronaut movement that would be popular twenty years later. Jessup believed that the pygmy races are millions of years older than scientists believe; long ago they developed an advanced technology which took them to the moon and beyond. These "little people," who figure in folklore and in UFO sightings, continue to observe us (also see entry: Moon Oddities).

Depressed by personal problems, Jessup dropped out of sight in 1958 and committed suicide a year later. Because of his involvement in the Allende hoax, tales began to arise that government agents killed Jessup because he knew too much.

said to have fallen into another dimension, where they interacted with aliens!

The Mysterious Carlos Allende

And who was the mysterious Carlos Miguel Allende? Some thought he might be an extraterrestrial himself. In actuality, he was an eccentric drifter, born Carl Meredith Allen in Pennsylvania in 1925. His wide reading in fantastic literature led him to develop his own strange ideas. Family members described him as a "master leg-puller" and recalled letters in which he all but admitted inventing the Philadelphia experiment. He also told them that he alone had written the notes in the Jessup book sent to the ONR.

Despite this damning information, the tale of the Philadelphia experiment lives on. Alfred Bielek, for example, lectures to New Age groups on his experiences during and after the event; he claims that he and his brother were aboard the ship—the U.S.S. *Eldridge*—when the experiment went awry. Bielek says that he and his brother jumped overboard, trying to escape, but found themselves hurled through "what seemed like a tunnel" into 1983. He also claims that later they met up with one of the mathematicians who designed the experiment. According to Bielek, he sent them back to 1943 just long enough for them to "destroy the equipment on board because it had created a time warp around the ship that could possibly engulf the planet."

REEL LIFE

The Philadelphia Experiment, 1984.

A World War II sailor falls through a hole in time and lands in the mid-1980s, whereupon he woos an attractive woman.

Philadelphia Experiment 2, 1993.

Melodramatic sci-fi thriller has Germany winning World War II by dropping a bomb on Washington. Southern California becomes one big labor camp with an evil mad scientist and a beleaguered hero who must time-travel back to 1943 to prevent the Germans from dropping that bomb.

Sources:

Genzlinger, Anna Lykins, *The Jessup Dimension,* Clarksburg, West Virginia: Saucerian Press, 1981.

Jessup, M. K., *The Case for the UFO,* New York: The Citadel Press, 1955.

Moore, William L., with Charles Berlitz, *The Philadelphia Experiment: Project Invisibility— An Account of a Search for a Secret Navy Wartime Project That May Have Succeeded— Too Well,* New York: Grosset and Dunlap, 1979.

Vallee, Jacques, *Revelations: Alien Contact and Human Deception,* New York: Ballantine Books, 1991.

AREA 51

Area 51 is located in a corner of the U.S. Nevada Test Site, where top-secret national security projects have been developed for several decades. These have included spy planes like the U-2 and the SR-71, the Stealth bomber, and the technology behind "Star Wars," or the Strategic Defense Initiative.

In recent years many observers have reported seeing odd lights moving in a manner that ordinary aircraft do not and cannot: flying at great speeds, stopping suddenly, and hovering for periods of time. Early sightings were almost all at night, but in a few cases witnesses saw the objects reflected in the moonlight and were able to tell that they resembled huge triangles. In May 1990 observers reported seeing such a craft in the daytime as well!

Writing in the October 1, 1990, issue of *Aviation Week & Space Technology,* John D. Morrocco reported that such triangles had been spotted not only in central Nevada but in parts of California. It appeared to him that a "quantum leap in aviation" had taken place, under conditions of great secrecy.

There had been similar sightings—as well as films and radar trackings—of giant triangles over Belgium and other European countries. These were usually considered UFO sightings. Yet to some, the Nevada/California reports and European accounts shared a link: the belief that the technology responsible for the remarkable new aircraft was from an unearthly source.

The Strategic Defense Initiative, commonly known as "Star Wars," is a space-based defense system that utilizes a layered weapon shield to track and destroy nuclear missiles heading for the United States—launched either from submarines or across continents—in any stage of flight. In March 1983 President Ronald Reagan called for a $30 billion, five-year program to research and develop the system, which combines several advanced technologies.

Outer Space Technology

Several close observers of the phenomenon thought that the breakthrough had come from studies of crashed extraterrestrial spacecraft stored at Area 51 of the Nevada test site. Rumors of this became so strong that hardly a magazine or newspaper account of the site failed to mention the notion. After investigating the stories, aircraft writer James C. Goodall reported, "Rumor has it that some of these systems involve force field technology, gravity drive systems, and 'flying saucer' designs. Rumor further has it that these designs are not necessarily of

Earth human origin, but of who might have designed them or helped us do it, there is less talk."

Other writers and researchers interviewed people who described secret projects based on extraterrestrial technology, but none of these informants were able to supply any proof. The one who attracted the most attention was Robert Scott Lazar, who, in November 1989, revealed startling information on a Las Vegas television news show. Lazar claimed that while working at Area 51 he learned that the advanced propulsion systems under development there involved almost unimaginable technology, powered by what he described as an "anti-gravity reactor." He also said that he had seen the crashed UFOs from which these technological secrets had come. According to Lazar, not even Congress knew about the project, and it was unlikely that the American public was meant to either.

Lazar's report caused a stir, but an investigation into his background—it was suggested that he was a liar and a lawbreaker—cast heavy doubt on his reliability. Still, the rumors surrounding Area 51 continued. And wild tales sprang up about government-extraterrestrial contact; one involved a treaty in which aliens supplied the American government with their technology in exchange for permission to kidnap its citizens!

One wild tale sprang up about a treaty in which aliens supplied the American government with their technology in exchange for permission to kidnap its citizens!

Sources:

Cooper, Milton William, *The Secret Government: The Origin, Identity, and Purpose of MJ-12,* Fullerton, California: The Author, May 23, 1989.

Good, Timothy, *Alien Liaison: The Ultimate Secret,* London: Random Century, 1991.

"Multiple Sightings of Secret Aircraft Hint at New Propulsion, Airframe Designs," *Aviation Week & Space Technology,* October 1, 1990, pp. 22-23.

Vanishing Acts

- THE BERMUDA TRIANGLE AND FLIGHT 19

- DEVIL'S SEA

- VILE VORTICES

Vanishing Acts

THE BERMUDA TRIANGLE AND FLIGHT 19

The term "Bermuda Triangle" was coined in 1964 by researcher Vincent Gaddis to describe an area in the Atlantic Ocean roughly bounded by Puerto Rico, the Bahamas, and the tip of Florida. Since the 1940s, several dozen ships and planes have disappeared in this area. Gaddis and many other writers have suggested that the Bermuda Triangle spelled doom for many who ventured into its domain, and many have speculated about "time warps," UFO kidnappings, and other paranormal reasons for the disappearances.

Of all the "mysterious disappearances" connected to the Bermuda Triangle, none is more famous than Flight 19. As with many of the stories behind the Triangle legend, however, research has demonstrated a serious gap between what has been claimed and what actually occurred in the ill-fated flight.

The Flight 19 Tragedy

At 2:10 on the afternoon of December 5, 1945, five Avenger torpedo bombers left the Naval Air Station at Fort Lauderdale, Florida, and headed east. Flight 19 was comprised of 14 men, all students in the last stages of training except for flight leader Charles Taylor. Taylor knew the Florida Keys well, but he did not know the Bahamas.

They were to conduct a practice bombing at Hens and Chicken Shoals, 56 miles away. Once that was done, the Avengers were to continue eastward for another 67 miles, then head north 73 miles. After that they would turn west-southwest and take the remaining 120 miles straight home. In short, they were flying a triangular flight path through what would later be called the Bermuda Triangle.

At 3:40 P.M. pilot and flight instructor Lieutenant Robert Cox, who was about to land at Fort Lauderdale, overheard a radio transmission among the Avenger bombers indicating that they were lost. A few minutes later he was able to contact Taylor, who told Cox that his compasses were not working. Because Taylor said he was sure he was over the Florida Keys, Cox urged him to fly north toward Miami.

Taylor was not, however, in the Keys. He was in the Bahamas. By flying north he would only go further out to sea. Cox and others (like the Port Everglades Boat Facility, an Air Sea Rescue Unit based near Forth Lauderdale) tried to pinpoint the location of Flight 19 but could not because of poor radio communications. Taylor was urged to turn over control of the flight to one of his students, which he did not do; overheard were arguments between him and other Flight 19 pilots, who thought that the group should fly west. Had they done so, they would have been saved.

By 4:45 P.M. observers on the ground were seriously concerned. It was clear from further messages from Taylor that, far from being temporarily lost, which happens to many pilots, he had no idea where he was. As dusk approached, radio transmissions grew fainter. Finally, at 5:15 Taylor radioed to Port Everglades that they were at last heading west. But fuel was a problem now; Taylor told his companions that as soon as one of them ran out of fuel they would all go down together.

The sun set at 5:29 P.M. With bad weather moving in from the north, the situation was growing ever more critical. Still, no one on the ground knew where Flight 19 was. At one point, Taylor was urged to switch to an emergency radio frequency, but he refused to do so for fear that he and the other planes would lose contact with one another.

By 5:50 P.M. the ComGulf Sea Frontier Evaluation Center thought it had determined the flight's position: east of New Smyrna Beach, Florida, and far to the north of the Bahamas. Soon after, however, Taylor was overheard ordering his other pilots to "turn around and go east again," explaining, "I think we would have a better chance of being picked up." Mistakenly, he still believed his group was over the Gulf of Mexico.

Because Flight 19's position had been so uncertain, no rescue aircraft had yet gone out. But at last a Dumbo flying boat (seaplane) left

But fuel was a problem now; Taylor told his companions that as soon as one of them ran out of fuel they would all go down together.

A group of Avenger torpedo bombers like Flight 19, which vanished December 5, 1945.

Miami at 6:20 P.M., heading northeast in an effort to reestablish contact with the lost flight. The Dumbo itself soon fell out of contact with shore, however, and for a while it was feared that it, too, was lost. The problem turned out to be ice on the radio antenna, and the seaplane continued on what would prove to be an unproductive search.

Within the hour, other aircraft, including two Martin Mariners, joined in the Dumbo's search. Mariner Trainer 32 had taken off around 7 P.M. east of New Smyrna Beach. Trainer 49 left the Banana River Naval Air Station some 20 minutes later and was to join up with the first Mariner. Lieutenant Gerald Bammerlin, 32's pilot, later told naval investigators, "When we arrived in the area of Flight 19's 5:50 position fix, about 8:15, ... the air was very turbulent and the sea very rough. We flew manually on instruments throughout the night."

In the meantime, Mariner 49 had failed to make its scheduled meeting. At 7:50 P.M. the crew of the S.S. *Gaines Mill* observed an enormous sheet of fire caused by the explosion of an airplane. A few minutes later the ship passed through a big pool of oil and looked without success for survivors or bodies. Though they saw some debris, crew members did not retrieve any of it because of the rough ocean. Weather conditions were worsening quickly.

By now it was certain that the Flight 19 aircraft had run out of fuel and were down. Taylor's last radio transmission was heard at 7:04 P.M. The search for the planes continued through the night, though with difficulty because of the high winds and raging sea. The next day hundreds of planes and ships looked, without success, for the missing Avengers and Mariner. No trace of them has ever turned up.

THE HOLLYWOOD RETURN OF FLIGHT 19

At the conclusion of Steven Spielberg's 1977 science-fiction film *Close Encounters of the Third Kind,* a UFO returns the Flight 19 crew to earth!

The Investigation

On April 3, 1946, at the end of an intense investigation of this much-discussed air disaster, the navy placed the blame on Taylor. According to investigators, the "flight leader's false assurance of identifying as the Florida Keys, islands he sighted, plagued his future decisions and confused his reasoning.... [He] was directing his flight to fly east ... even though he was undoubtedly east of Florida." When Taylor's mother and aunt refused to accept this verdict, the navy set up a panel to review the report. In August this panel announced it could only agree with the first conclusion. Furious, the two women hired a lawyer for a hearing the following October. On November 19 the Board for Correction of Naval Records retracted the original verdict and officially blamed the disaster on "causes or reasons unknown."

The fate of Mariner 49, however, seemed clear. The Mariners were called "flying gas bombs" because of the dangerous fumes they were known to emit. Something as small as a lighted cigarette or an electrical spark could ignite them. As for the Avengers, the 50-foot-high waves tearing across the ocean surface had probably chewed them up and sent what remained to the bottom in a matter of seconds.

Although Flight 19—along with everything else connected with the Bermuda Triangle—became a sensation from the 1950s through the 1970s, later writers examined the event carefully, some digging even

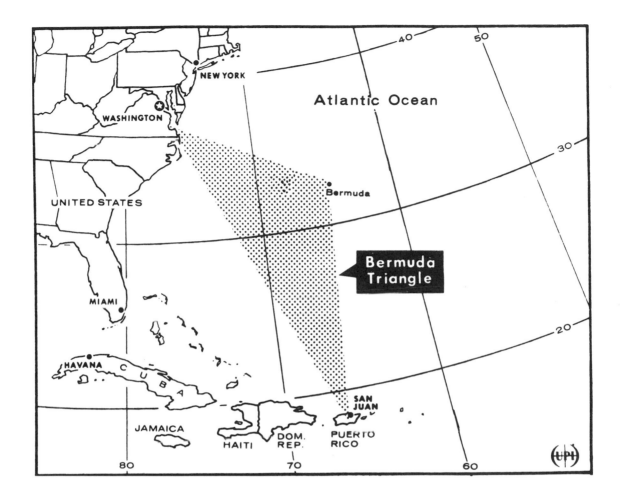

The Bermuda Triangle.

deeper than the navy's original investigation. In 1980 Larry Kusche, whose meticulous research debunked the Bermuda Triangle and "devil's sea" phenomena, published *The Disappearance of Flight 19.* Kusche felt that the navy correction board should not have excused Taylor. Though the "decision was a kindness to Mrs. Taylor [Charles's mother] ... it was incorrect. The conclusion of the original Board of Investigation, that Charles Taylor was at fault, was correct."

Reflecting on the mission of Flight 19 in 1985, Willard Stoll, who had taken off with Flight 18 half an hour before Taylor's mission, remarked: "What the hell happened to Charlie? Well, they didn't call those planes 'Iron Birds' for nothing. They weighed 14,000 pounds empty. So when they ditched, they went down pretty fast. But they found the *Titanic,* and maybe one day they'll find him and the others. Wherever they are, they're together."

Mystery and Myth

In September 1950 Associated Press reporter E. V. W. Jones sent a story out over the wires. In it he wrote that the same triangular area connecting Florida, Bermuda, and Puerto Rico in which Flight 19 disappeared was a "limbo of the lost" where planes and ships often "vanished in the thin air." Especially baffling, he noted, was the disappearance of Flight 19 and the Martin Mariner that had gone in search of it. An October 1952 article in *Fate,* a popular magazine devoted to "true mysteries," was largely based on the Associated Press piece; it discussed the case of Flight 19, as well as other disappearances.

In the 1955 book *The Flying Saucer Conspiracy,* Donald E. Keyhoe, a retired marine corps major and believer in extraterrestrial visitors, suggested that a "giant mother ship" from space had snatched the planes of Flight 19. Like many other writers who would follow, Keyhoe claimed that the sea had been calm that day. More influential, however, was an *American Legion* magazine article written by Allan W. Eckert. He concocted dialogue that other Triangle writers would reprint again and again. According to Eckert, Taylor had radioed Fort Lauderdale that "everything is wrong ... strange ... the ocean doesn't look as it should."

A later writer, Art Ford, reported that he had interviewed a radio operator who had heard Taylor say, "They look like they're from outer space—don't come after me." Nothing in the records of Taylor's conversations during the flight supported that claim.

In a February 1964 *Argosy* piece, and the next year in the book *Invisible Horizons,* Vincent Gaddis called Flight 19's case the "most incredible mystery in the history of aviation." Other authors would report equally fantastic and misleading versions of the episode. Rejecting all possible ordinary explanations for the tragedy, they seized on aliens, the fourth dimension, space-time travel, and mysterious magnetic forces.

A FALSE ALARM

In 1991 a crew of the ship *Deep See,* hunting for sunken Spanish galleons (sailing ships), found the intact remains of five Avengers on the ocean floor, 600 feet down and 10 miles northeast of Fort Lauderdale. One plane bore the number 28, the same as Taylor's aircraft. Further investigation revealed, however, that the craft were not from Flight 19 after all; the numbers on the other planes did not correspond with those of the tragic flight. In addition, the craft were older-model Avengers.

Others Vanish in the Triangle

In February 1963 the *Sulphur Queen,* a tanker carrying 15,260 tons of molten sulphur, mysteriously disappeared as it neared the Straits of Florida. After an extensive air and sea search, no bodies, lifeboats, or

oil slicks were found. Many people believed that the ship was forever lost in the Bermuda Triangle.

Eventually, however, some wreckage did surface, including a piece of an oar, a name board with the letters "ARINE SULPH," and a life preserver and life jacket that bore the ship's name. Investigators speculated that the ship (which was structurally weak) could have encountered bad weather and either blown up due to its flammable cargo or simply sank into the deep water.

Still another Triangle mystery occurred in December 1967, when a 23-foot cabin cruiser called *Witchcraft* disappeared with its two-man crew just off Miami Beach. The men reported a damaged propeller to the Coast Guard and asked to be towed back to port. They also reported that the boat's hull was intact and that the built-in flotation chambers rendered the craft virtually unsinkable. Yet when the Coast Guard reached the location—after only nineteen minutes had passed—there was no trace of the boat, the men, or their life preservers.

Bermuda Triangle Writings Abound

In his 1955 book *The Case for the UFO,* M. K. Jessup suggested that aliens were responsible for the Bermuda Triangle disappearances, a view shared by authors Donald E. Keyhoe (*The Flying Saucer Conspiracy,* 1955) and Frank Edwards (*Stranger Than Science,* 1959). In "The Deadly Bermuda Triangle," an *Argosy* article published in February 1964, Vincent H. Gaddis introduced the catch-all phrase "Bermuda Triangle," which was forever associated with the phenomenon. Soon nearly every book on "true mysteries" discussed the Bermuda Triangle or, as some called it, the "devil's triangle" or "hoodoo sea." Ivan T. Sanderson, author of *Invisible Residents* (1970), felt that an intelligent, technologically advanced underwater civilization was behind the Triangle disappearances, as well as other mysterious occurrences like UFOs.

The first book devoted solely to the Triangle, John Wallace Spencer's 1969 *Limbo of the Lost,* enjoyed a huge readership when it was published as a paperback in 1973. In 1970 a feature-film documentary, *The Devil's Triangle,* brought the subject to a new, larger audience. Public interest peaked in 1974 with the publication of *The Bermuda Triangle,* written by Charles Berlitz with J. Manson Valentine; it sold five million copies worldwide. That same year two paperbacks, Richard Winer's *The Devil's Triangle* and John Wallace Spencer's *No Earthly Explanation,* also sold extremely well.

But readers began to notice that Triangle authors seemed to be rewriting each other's work, their articles and books indicating little evidence of original research. In 1975 Larry Kusche, a librarian at Arizona State University, published *The Bermuda Triangle Mystery— Solved,* a book based on detailed information other authors had failed to unearth. Weather records, reports of official investigative agencies, newspaper accounts, and other documents showed that Triangle writers had played fast and loose with the evidence. For example, calm seas in their writings turned out to be raging storms in reality; mysterious disappearances were sinkings and crashes with ordinary causes; the remains of ships "never heard from again" had really been found long before.

In an April 4, 1975, letter to *Fate* magazine editor Mary Margaret Fuller, a spokesman for Lloyd's of London, an international insurance corporation known for its underwriting of shipping, wrote: "According to Lloyd's Records, 428 vessels have been reported missing throughout the world since 1955, and it may interest you to know that our intelligence service can find no evidence to support the claim that the 'Bermuda Triangle' has more losses than elsewhere. This finding is upheld by the United States Coastguard [sic] whose computer-based records of casualties in the Atlantic go back to 1958."

Triangle believers could not defend their position in light of these findings. The Bermuda Triangle had turned out to be—in Kusche's words—a "manufactured mystery." While articles on the subject still appear in sensational publications from time to time, it is viewed by most as a fad that once had a stronghold in the collective imagination.

Sources:

Berlitz, Charles, with J. Manson Valentine, *The Bermuda Triangle,* Garden City, New York: Doubleday and Company, 1974.

Berlitz, Charles, with J. Manson Valentine, *Without a Trace,* Garden City, New York: Doubleday and Company, 1977.

Cazeau, Charles J., and Stuart D. Scott, Jr., *Exploring the Unknown: Great Mysteries Reexamined,* New York: Plenum Press, 1979.

Clary, Mike, "Mystery of 'Lost Patrol' May Be Solved," *Los Angeles Times,* May 18, 1991.

Gaddis, Vincent, *Invisible Horizons: True Mysteries of the Sea,* Philadelphia, Pennsylvania: Chilton Books, 1965.

Kusche, Larry, *The Disappearance of Flight 19,* New York: Harper and Row, 1980.

Kusche, Lawrence David, *The Bermuda Triangle Mystery—Solved,* New York: Harper and Row, 1975.

Winer, Richard, *The Devil's Triangle,* New York: Bantam Books, 1974.

DEVIL'S SEA

During the **Bermuda Triangle** fad of the 1970s, several writers maintained the existence of another mysterious area of lost ships and planes, this one off the eastern or southeastern coast of Japan. Called the "devil's sea," it was a place where disappearances were so sudden that affected craft usually had no time to send out distress signals before vanishing.

As with the Bermuda Triangle, it was theorized that unknown space-time, magnetic, or gravitational forces (or kidnappers from outer space) were behind the disappearances. It was also claimed that the Japanese government was extremely alarmed by the phenomenon.

The devil's sea was called a "true mystery" in articles and books on such subject matter until Larry Kusche, an Arizona State University librarian whose detailed research debunked the Bermuda Triangle, took a closer look at it. Kusche traced the origins of the devil's sea legend to *New York Times* stories published on September 27 and 30, 1952. They told of an unusual ocean disaster, the sinking of two Japanese ships by a tidal wave stemming from an underwater volcano. A January

Called the "devil's sea," it was a place where disappearances were so sudden that affected craft usually had no time to send out distress signals before vanishing.

15, 1955, *Times* report of another ship disaster in the area used the term "devil's sea" and called it the "mystery graveyard of nine ships in the last five years."

In the early 1970s Kusche corresponded at length with officials from Japan and nearby islands. None had ever heard of the "devil's sea," and all insisted that sinkings in the area were not mysterious or inexplicable. Indeed, the absence of radio messages in some cases from the early 1950s was easily explained: poor owners of smaller fishing vessels could not afford radios.

Writing about the devil's sea in 1975, Kusche concluded, "The story is based on nothing more than the loss of a few fishing boats 20 years ago in a 750-mile stretch of ocean over a period of five years. The tale has been reported so many times that it has come to be accepted as fact."

Sources:

Kusche, Larry, *The Bermuda Triangle Mystery—Solved,* Buffalo, New York: Prometheus Books, 1986.

Nichols, Elizabeth, *The Devil's Sea,* New York: Award Books, 1975.

Sanderson, Ivan T., *Invisible Residents: A Disquisition upon Certain Matters Maritime, and the Possibility of Intelligent Life Under the Waters of This Earth,* New York: World Publishing Company, 1970.

VILE VORTICES

Ivan T. Sanderson was a zoologist who made a career of studying all kinds of strange physical events. Sanderson believed that an intelligent civilization lived in the oceans of the earth. He suspected that its members, whom he called "Other Intelligences" or OINTS, came from "elsewhere" and had settled in the ocean to avoid detection. At least some UFOs (also see entry: **Unidentified Flying Objects**), Sanderson felt, were this civilization's aircraft, which doubled as submarines.

Noting that planes and ships seemed to be disappearing mysteriously, Sanderson suggested that these ocean OINTS might be responsible. According to Sanderson, the **Bermuda Triangle** was one of ten diamond-shaped areas stretching in parallel bands above and below the equator in which the OINTS operated. Sanderson used the term "vile vortices" for these areas. (A vortex is a body of water or other fluid that runs in a whirling, circular motion, like a whirlpool, forming a vacuum that draws all things to its center.) The two polar regions were also included in Sanderson's list of vile vortices.

Sanderson pointed out that, aside from mysterious plane, ship, and submarine disappearances, an unusual number of UFO sightings had been reported in these areas. So had strange weather occurrences—like sudden high winds, storms, and rough water. He suspected that OINTS were forced to take dramatic action at times to keep their presence a secret. This might even require seizing a "whole ship and everything in and on it."

Sanderson's evidence of the existence of vile vortices was feeble at best; in some cases it hinged on shadowy rumors of unexplained events in the areas in question. Even in the case of the most famous and best-studied "vortice" (vortex)—the Bermuda Triangle—detailed research by Larry Kusche and others exposed the disappearances as a "manufactured mystery." Weather records, official investigations, newspaper accounts, and other documents revealed that ordinary causes were behind the sinkings and crashes that occurred there.

Sanderson believed that an intelligent civilization lived in the oceans of the earth.

Sources:

Kusche, Larry, *The Bermuda Triangle Mystery—Solved,* Buffalo, New York: Prometheus Books, 1986.
Sanderson, Ivan T., *A Disquisition upon Certain Matters Maritime, and the Possibility of Intelligent Life Under the Waters of the Earth,* New York: World Publishing Company, 1970.
Sanderson, Ivan T., *More "Things,"* New York: Pyramid Books, 1969.

Light Shows

- GHOST LIGHTS
- GREEN FIREBALLS
- BROWN MOUNTAIN LIGHTS
- MARFA LIGHTS

Light Shows

GHOST LIGHTS

Ghost lights are usually luminous, glowing points or spheres of light the appearance, behavior, or location of which puts them in a different category from **ball lightning** or **unidentified flying objects.** Ghost lights have often been considered supernatural, with their appearances signaling coming death or earthly visits from those who have already departed. Where ghost lights have appeared regularly in one place over a period of time—as with the famous **Brown Mountain lights** or the **Marfa lights**—legends have sprung up around them.

Lights in Folk Tradition

Over three hundred years ago, in the book *The English Empire in America* (1685), Nathaniel Crouch reported how he once witnessed the mysterious flame of which the Indians spoke, which would appear at night before the wigwams of those who were soon to die.

Three decades earlier, in 1656, John Davis, the vicar of Geneu'r Glyn, Cardiganshire, Wales, had also reported how he and others had observed colored lights that seemed to predict death. The lights could appear anywhere: in the open air, on their way through a door, or inside a house. A small light signaled the death of a child, a bigger light that of an adult. Several lights together meant as many deaths. A relative, Davis related, had once seen five lights in a room, and that very night five servants suffocated to death in a freak accident there.

In 1897 R. C. Maclagan published a long survey of ghost-light traditions, stories, and reports from Scotland's West Highlands.

Many of them told how lights appeared on the rocky shores there, predicting exactly where shipwrecks and boating accidents—and the resulting drownings—would occur. Accounts of these "corpse candles," as the lights were sometimes called, continued into the twentieth century.

One Welsh witness offered folk scholar W. Y. Evans-Wentz these observations:

> The death-candle appears like a patch of bright light; and no matter how dark the room or place is, everything in it is as clear as day. The candle is not a flame, but a luminous mass, lightish blue in color, which dances as though borne by an invisible agency [agent], and sometimes it rolls over and over. If you go up to the light, it is nothing, for it is a spirit.

What was believed to be a "corpse candle" was sighted in the mountains of Stockton, Pennsylvania, in February 1909. It caused great excitement among local residents and was the subject of several newspapers accounts. One described the "appearance at night of an arrow of flame, which hovers over the spot on the mountain where the dismembered body of a woman was found in a barrel two years ago.... The light appears every night at about 9 o'clock and hovers over the spot until midnight, but it disappears when anyone approaches the spot to investigate." The account added that some superstitious villagers believed that the light was the spirit of the slain woman, trying to keep the memory of the crime alive so that her murderers would someday be caught.

Lights have also been associated in folk tradition with appearances of **fairies.** A young Irishman who attended Oxford University with Evans-Wentz told the folk scholar about his strange experience in the winter of 1910. He and a companion were on their way home from Limerick on horseback when they noted a light in the distance "moving up and down, to and fro, diminishing to a spark, then expanding into a yellow luminous flame." Later the travelers noticed two lights similar to the first they had seen, these expanding into flames "about six feet high by four feet broad. In the midst of each flame we saw a radiant being having human form." The lights moved toward one another and when they made contact the two men could see the beings walking side by side, their bodies "formed of a pure dazzling radiance, white like the radiance of the sun, and much brighter than the yellow light or aura surrounding them." So dazzling were the haloes that surrounded the beings' heads, in fact, that the observers were unable to make out their faces.

Religious Lights

In early December 1904, a 38-year-old Welsh housewife and folk preacher, Mary Jones of Egryn, Merionethshire, reported seeing a vision of Jesus. Jones quickly became the leading figure in a Christian revival that, in the weeks and months ahead, attracted international attention. While the evangelist's message about Christian ideals was not unusual, the strange lights that were often seen when she preached were! Odder still, the lights were visible to some people—including many doubting journalists—but not to others.

A *London Daily Mirror* reporter related a sighting he witnessed while in the company of the newspaper's photographer; they were stationed in Egryn one evening, hoping to see the lights. After a three-and-a-half-hour wait, a light resembling an "unusually brilliant carriage lamp" appeared not far from the chapel where Jones led her ministry. As the reporter walked nearer, "it took the form of a bar of light quite four feet wide," he wrote, and "a kind of quivering radiance flashed with lightning speed from one end of the bar to the other" before it disappeared. While two women near the reporter also marveled over the light, he was astonished to find that a nearby group of some 15 or 20 people had seen nothing at all!

Beriah G. Evans of the *Barmouth Advertiser* also wrote of his strange experience with the lights. While walking with the preacher Jones and three others early in the evening of January 31, 1905, he saw "three brilliant rays of light strike across the road from mountain to sea.... There was not a living soul there, nor house, from which it could have come." Half a mile later, a "blood-red light" appeared immediately before them, a foot above the ground, in the middle of a village street. Only the reporter and Jones witnessed these phenomena. Evans wrote of another London journalist who experienced a similar event soon afterward: the man and his companion saw a broad band of white light on the road and walls near the chapel, but a group of a half dozen other people there detected nothing at all.

Sightings of these lights were plentiful, with many separate and multiple witnesses. Once, as Jones was holding a revival meeting in a chapel in Bryncrug, a ball of fire cast rays downward, lighting up the church. On another occasion, Jones and three companions were traveling in a carriage in broad daylight when a bright light with no visible source suddenly shined on them. The riders of two trailing carriages, including a pair of skeptical journalists, witnessed the sight, as did Barmouth residents awaiting the preacher's arrival. Some accounts of the lights included supernatural or spiritual elements.

> Odder still, the lights were visible to some people—including many doubting journalists—but not to others.

On February 23 the *Advertiser* noted that two men—one an important farmer—reported a "gigantic human form rising over a hedgerow. Then a ball of fire appeared above and a long ray of light pierced the figure, which vanished."

During all this, Jones and some of her followers were also visited by Christ and angels, through dreams or visions. And one dark night, as she walked along a country road, Jones said she met a shadowy figure who turned into a **black dog,** which charged her and was driven off only after she began to sing a hymn. She believed the attacker was Satan. While such religious experiences are highly subjective, open to all sorts of questions and explanations and thus regularly discounted, these lights, at least, have been the source of serious research for nearly a century. The appearance of the lights may or may not have been related to the religious revival led by Mary Jones, for history shows that most accounts of strange lights do not have religious connections.

Local Lights

Vincent H. Gaddis, who writes about mysterious events, felt that in hundreds, possibly even thousands, of places around the world, "strange lights haunt the earth." Unlike UFOs or ball lightning, "they are usually small in size and appear close to the ground," he described, with "their outstanding characteristic [being] that they are localized to one area or place." Legends often grow up around these frequent lights; a common one describes them as lanterns carried by ghosts who are searching for something they lost in life, including—in the scariest cases—their heads!

Few accounts of strange lights are actually investigated by scientists or serious researchers, and when they are the results are usually disappointing. Many lights, for example, turn out to be from headlights of cars on distant highways or from stars and planets the light from which has been refracted, or bent, through layers of air of different temperatures. Still, there are some cases that have scientists and trained observers baffled. Two such examples are the lights at the Yakima Indian Reservation of south-central Washington and those frequenting the Hessdalen Valley of Norway.

Yakima

The Yakima reservation is a thinly populated area 3,500 miles square and divided between rugged wilderness in the west and flatlands in the east. Beginning in the late 1960s (though a few sightings had occurred before then), forest rangers, fire-control workers, and others began reporting the movement of bright white lights low in the sky. When chief fire-control officer W. J. (Bill) Vogel heard these reports, he judged them unimportant and even foolish—that is, until he experienced some strange sightings of his own!

Late one night when he was on patrol, he saw a luminous object that had a teardrop shape, with the small end pointing up. It was "brilliantly white in the center," surrounded by a fluorescent tan or light orange halo. Vogel reported that the object's "most awe-inspiring feature was a mouselike tail or antenna protruding from the small end and pointing upward. The antenna, as long as the object itself, was segmented into colors of red, blue, green, and white which were constantly changing brilliancy and hue."

Over the next 90 minutes Vogel took a series of photographs of the object, which eventually vanished over the mountains. It would be the first of many sightings he would report. Soon the fire-control chief was busy collecting and investigating reports made by others, most of them from his own highly trained and reliable fire lookouts. And many local people also had stories to tell.

Investigators came to Yakima, including astronomer and former air force UFO adviser J. Allen Hynek. He persuaded the Indian Tribal Council to allow an observer to set up equipment on the reservation to note the lights' activity. The observer, David Akers of the Aerial Phenomena Research Organization (APRO), brought with him cameras and other recording devices. On August 19, 1972, his first night on the reservation, Akers, along with Vogel, saw two round, glowing, reddish-orange lights circling, changing places, and flashing on and off. He took four photographs. Other sightings and more photographs followed until Akers left the reservation at the end of the month.

While equipment difficulties kept Akers from obtaining some of the data he was seeking, he left the reservation convinced that "something very strange and unusual is taking place." He returned over the next few days to interview witnesses and to view and photograph more strange lights. The detailed records that he and Vogel kept showed that the lights appeared at ground level, above ground level, and at high altitudes. Observer Greg Long would later join the investigation and write a book about his findings.

Some of the oddest experiences reported at Yakima by fire lookouts involved a sort of mental communication with the lights. While most sightings involved distant lights, lookouts sometimes were able to view them from several hundred yards. Still, they could not get any closer: lookouts reported "hearing" a voice inside their heads warning, "Stay back, or you'll get hurt," and feeling restrained.

One lookout saw a shaft of bright purple light shining down around her cabin. When she tried to go outside to investigate, she felt as if "two magnets [were] repelling each other" and blocking her exit. Puzzled but determined, she ran at the doorway several times but could not get through. Weirder still, some observers of the lights reported feeling as if they were seeing something that they were not meant to see and removed themselves altogether from the objects they had originally set out to investigate.

It should also be noted that a handful of the Yakima reports included sightings of spacecraft and alien beings. So, while the lights there seemed to behave like true ghost lights, they may, in fact, have had some connection to UFOs. In any case, the number of sightings at the Yakima Indian Reservation fell off to a large degree after 1986.

Hessdalen

The Hessdalen lights also subsided in 1986, but for a period of several years they were the target of an all-out investigation by ufologists, scientists, and local residents. The Hessdalen Valley, stretching across 12 kilometers (about 7 1/2 miles) of central Norway near the Swedish border and having no more than 150 inhabitants, began to experience odd light shows in November 1981.

The lights sometimes appeared as often as four times a day, often along mountain-tops, near the ground, or on the roofs of houses. Usually white or yellow-white, they typically were shaped like cigars, spheres, or an "upside-down Christmas tree."

In one case of an upside down Christmas tree shape, the light, according to miner Bjarne Lillevold, was "bigger than the cottage beside it. It was about four meters above the hill and had a red blinking light on it.... The object moved up and down like a yo-yo for about 20 minutes. When it was close to the ground, the light faded, but at the height of the maneuver it was so bright that I could not look at it for long. When the light was near the ground, I could see through it as though it was made of glass."

Project Hessdalen

Once in a while, according to other witnesses, a red light held a position in front. The lights hovered, sometimes for an hour, then shot off at extraordinary speed. Most of the time they traveled from north to south. Investigators from UFO-Norway brought valley residents together to discuss their sightings on March 26, 1982. Of the 130 who attended, 35 said they had seen the lights.

In the summer of 1983 Scandinavian ufologists established Project Hessdalen, an investigative force that included scientists from the Universities of Oslo and Bergen. A variety of equipment was set up on three mountains. The month-long winter watch, from January 21 to February 26, 1984, produced some sightings, radar trackings, and photographs that proved interesting but settled nothing. Most exciting, perhaps, was the fact that the lights seemed to respond when laser beams were aimed at them. Once, on February 12, for example, the lights changed their single flashes to a double-flash pattern—in a way that seemed almost knowing—whenever investigators used the laser.

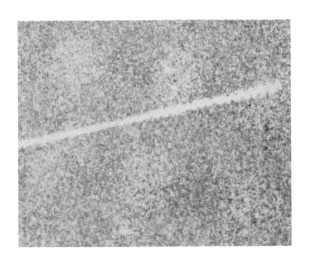

This 10-second time exposure was taken by Roar Wister on the evening of February 21, 1984, in Norway's Hessdalen Valley.

Ultimately, investigators disagreed on what the lights could be, with some convinced they were a reflection of some—as yet unknown—geophysical event (occurring beneath the earth's surface), others feeling that they showed a guiding intelligence. The strange "coincidences" that took place throughout the investigation also seemed to point to an intelligence behind the lights.

Investigator Leif Havik wrote: "On four separate occasions, it happened that we came to the top of Varuskjolen, stopped the car, went outside, and there 'it' came immediately and passed by us. The same thing happened once on Aspaskjolen. All these instances happened at different times of the day and most of the time it was an impulse which made us take video equipment which recorded the radar screen. One evening the pen of the magnetograph failed to work. At the same time the video tape had come to an end, and the phenomenon appeared less than one minute later. The next evening we made certain that the pen had sufficient ink and turned on the video recorder ten minutes later than the night before. We thought then ... everything was ready for the usual 10:47 'message.' [One light appeared regularly at 10:47 P.M.] The video tape ran out at 10:57 P.M. and we thought that tonight 'it' had failed us. But at 10:58 the usual phenomenon appeared."

Nonetheless, in terms of reliable scientific data, Project Hessdalen was a disappointment. Investigators logged 188 sightings. Some of

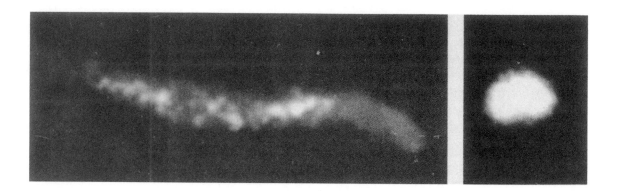

These two Hessdalen photographs were snapped just a few seconds apart, showing the light's abrupt change in shape.

these were passing aircraft, and most photographs proved too unclear to study. Looking back on the investigation, University of Oslo physicist Elvand Thrane, a member of the project, remarked, "I'm sure that the lights were real. It's a pity we cannot explain them."

Explaining Ghost Lights

Unexplained lights in the sky are probably caused by a number of things, from the ridiculously common to the truly mysterious. Two interesting ideas, Paul Devereux's "earthlights" and Michael Persinger's "tectonic stress theory," suggest that sightings of strange lights are caused by geophysical activity. While rejected by most scientists, both theories hold that ghost lights are produced by subterranean (underground) processes that not only create luminous energy on the earth's surface, but may cause hallucinations in its observers as well!

Sources:

Devereux, Paul, *Earth Lights Revelation: UFOs and Mystery Lightform Phenomena: The Earth's Secret Energy Force,* London: Blandford, 1989.
Gaddis, Vincent H., *Mysterious Fires and Lights,* New York: David McKay Company, 1967.
Long, Greg, *Examining the Earthlight Theory: The Yakima UFO Microcosm,* Chicago: J. Allen Hynek Center for UFO Studies, 1990.
McClenon, James, *Deviant Science: The Case of Parapsychology,* Philadelphia: University of Pennsylvania Press, 1984.

GREEN FIREBALLS

For a three-year period, between late 1948 and 1951, a flurry of sightings of "green fireballs" occurred in the southwestern United States. At one point military and other government agencies feared that enemy agents were somehow connected with the fireballs, which were spotted near some of America's most secret national-security bases.

Early Sightings

The first sightings to attract official attention took place on the evening of December 5, 1948, when two pilots flying over New Mexico reported two separate observations, 22 minutes apart, of a pale green light that was visible for no more than a few seconds. The witnesses insisted that it was not a meteor that they had seen, but some kind of strange flare. The next day a similar "greenish flare" was spotted for three seconds over the supersecret nuclear installation Sandia Base, part of the Kirtland Air Force Base complex in New Mexico.

Also on December 6, the Seventh District Air Force Office of Special Investigations (AFOSI) at Kirtland launched a study. And two pilot investigators observed one of the strange objects about 2,000 feet above their aircraft the following evening. They noted that it seemed to move parallel to the earth's surface and that it resembled the flares commonly used by the air force. "However, the light was much more intense and the object appeared to be considerably larger than a normal flare," they reported. "[It] was definitely larger and more brilliant than a shooting star, meteor, or flare." After only a couple of seconds, "the object seemed to burn out [and] ... a trail of glowing fragments reddish orange in color was observed falling toward the ground."

The next day one of the officers contacted Lincoln La Paz, director of the University of New Mexico's Institute of Meteoritics (meteor studies) and an air force consultant on top-secret matters. La Paz acknowledged that the green flares were different from any meteors he had ever heard of. Not long afterward, the scientist saw one of the lights himself. From his own observations and those of two other witnesses, from the Atomic Energy Security Service or AESS, La Paz decided that the object had flown too slowly and too silently to be a meteor. In a letter to the investigation's commanding officer, he noted that "none of the green fireballs has a train of sparks or a dust cloud.... This contrasts sharply with the behavior noted in cases of meteoritic fireballs—particularly those that penetrate to the very low levels where the green fireball of December 12 was observed."

> After only a couple of seconds, "the object seemed to burn out [and] ... a trail of glowing fragments reddish orange in color was observed falling toward the ground."

At La Paz's suggestion, the AESS organized patrols to try to photograph the fireballs. And scientists and engineers at New Mexico's Los Alamos Scientific Laboratory set up a group to study sighting reports. As the number of accounts continued to grow, the army and the air force became more and more concerned; they were especially disturbed when La Paz concluded, by early 1949, that the fireballs were not a natural occurrence, but had been put there by somebody or something.

METEOR

A meteor is one of the small pieces of matter in the solar system that can be seen only when it falls into the earth's atmosphere, where friction may cause it to burn or glow. When this happens it is sometimes called a "falling" or "shooting" star.

Military Scare?

On February 16 a Conference on Aerial Phenomena brought military officers and scientists to Los Alamos, where they were assured that the fireballs were not the result of any secret U.S. military operation. La Paz was eager to ask if any of the conference members knew of meteors that acted like the green fireballs—moving in long horizontal paths at a steady speed, eight to ten miles off the ground.

In late April Pentagon and air force officials sent physicist Joseph Kaplan to Kirtland. He, La Paz, and others discussed establishing a network of instruments and observers at several locations in New Mexico. Meanwhile, since early March, tiny white lights or "flares" had been spotted regularly near Killeen Base, a nuclear-weapons storage site inside Camp Hood in central Texas. This caused great alarm. Colonel Reid Lumsden, commander of AFOSI at Kelly Air Force Base in San Antonio, Texas, stated that the "unknown phenomena in the Camp Hood area could not be attributed to natural causes."

Still, despite the testimony of local experts and witnesses, officials in Washington decided that the fireballs and lights were natural occurrences, even if they did have features that—in Kaplan's words—were "difficult to explain." The sightings continued.

In the summer of 1949, samples of the New Mexico atmosphere were examined; they revealed an unusually large and unexplained amount of copper particles, suggesting a connection with the fireball sightings. La Paz viewed this as further evidence that the fireballs were not meteors: "I know of no case in which even the tiniest particle of copper has been reported in a dust collection supposedly of meteoritic origin," he maintained.

Project Twinkle

After meeting with high-ranking air force officials, Kaplan urged that a photographic and spectrographic patrol be set up to collect data on the fireballs and the lights. (A spectograph is an instrument that photographs and measures the display of light or other radiation of an object or substance by breaking it into a spectrum.) Thus Project Twinkle began, which consisted of two observers stationed at an operations post at Holloman AFB in New Mexico. La Paz thought the matter deserved a far more "intensive, systematic investigation" than this basic program and was very disappointed.

Despite some interesting sightings, Project Twinkle shut down in December 1951 because of problems with funding, instruments, workers, and officials. Many felt that the failure of the project was the loss of an opportunity to collect solid information on at least one kind of unidentified flying object. Many of the scientists who were involved in the investigation remained convinced that the fireballs were not natural, but artificially created. When Captain Edward J. Ruppelt, head of the air force's UFO investigation unit, discussed the subject with Los

The spectacular "green fireballs" observed in the southwestern United States in the 1940s and 1950s remain great puzzles to UFO buffs.

Alamos scientists in 1953, for example, they told him that they believed the objects were fired from extraterrestrial spacecraft.

Sources:

Clark, Jerome, *The Emergence of a Phenomenon: UFOs from the Beginning through 1959— The UFO Encyclopedia,* Volume 2, Detroit: Omnigraphics, 1992.

Ruppelt, Edward J., *The Report on Unidentified Flying Objects,* Garden City, New York: Doubleday and Company, 1956.

BROWN MOUNTAIN LIGHTS

Brown Mountain (altitude 2,600 feet) is located in the Blue Ridge Mountains of western North Carolina near Morganton. Many stories, and even songs—including Scotty Wiseman's popular bluegrass tune "The Brown Mountain Light"—have sprung up around the mysterious lights that have long appeared there.

Sightings

Writing in a 1925 issue of *Literary Digest,* Robert Sparks Walker reported that sightings of the lights varied greatly. One person described the light he observed as "pale white ... with a faint, irregular halo." This witness said that the light moved in a circle several times, disappeared, then returned and continued its circular motion. Another observer saw "a steady glowing ball of light" that was yellow in color. "Like a star from a bursting skyrocket," the witness reported, it lasted for about half a minute, then quickly disappeared. "To some people it appears stationary; to others, it moves sometimes upward, downward, or horizontally," Walker noted. "A minister says that it appeared like a ball of incandescent light in which he could observe a seething motion."

The first printed account of the lights appeared in the *Charlotte Daily Observer* on September 13, 1913. A group of fishermen had reported seeing a "very red" mystery light "just above the horizon almost every night." Not long afterwards, D. B. Sterrett of the U.S. Geological Survey investigated and concluded that train headlights were the cause. But members of a 1916 expedition to the area swore that they had seen the lights act in a very "unheadlightlike" way: floating in and out of gullies and ravines, for example.

Members of a 1916 expedition to the area swore that they had seen the lights act in a very "unheadlightlike" way: floating in and out of gullies and ravines, for example.

Blue Ridge Mountains in western North Carolina.

Fireflies?

More sightings—and arguments about their source—brought another Geological Survey scientist, George Rogers Mansfield, to the area in March and April 1922. He examined the mountains, interviewed local residents, and observed the lights for seven nights. He concluded that 44 percent of the light sightings were related to automobiles, 33 percent to trains, 10 percent to other, stationary lights, and 10 percent to brushfires—leaving just 3 percent of the accounts unexplained. He also thought that the 1916 report might have involved fireflies.

In the years since then, witnesses have reported more mystery lights, which have resembled "toy balloons," "misty spheres," "floodlights," and "sky rockets." A few times, when observers have gotten close to the lights, they have reported hearing a sizzling noise. A 1977 experiment beamed a powerful arc of light from a town 22 miles away to a location west of the mountain where observers waited. According to them, the blue-white beam looked like an "orange-red orb apparently hovering several degrees above Brown Mountain's crest."

This pretty much convinced investigators that the sightings were caused by the refraction (bending through layers of air) of distant lights. Still, the folklore of the Blue Ridge Mountains is rich with stories of people who witnessed the lights long before the age of trains and cars and electric lights!

Sources:

Devereux, Paul, *Earth Lights Revelation: UFOs and Mystery Lightform Phenomena: The Earth's Secret Energy Force,* London: Blandford Press, 1989.
Walker, Robert Sparks, "The Queer Lights of Brown Mountain," *Literary Digest,* November 7, 1925, pp. 48-49.

MARFA LIGHTS

They observed three lights this time; one seemed to be aware of their presence, almost daring them to chase it!

Marfa, a ranching town of about 2,400 people at the west end of southern Texas, seems like an ordinary place except for one thing: it is the regular site of ghost lights. While they do not show up every night, they do appear often and are visible to anyone who looks southwest toward the Chinati Mountains 50 miles away.

Sometimes the lights, which resemble train or automobile headlights, do little more then flicker. Sometimes they jump across the desert floor or split into new lights. At other times they rise lazily into the air. Many photographs have been taken of these remarkable lights.

Geologists Pat Kenney and Elwood Wright reported two extraordinary sightings of the lights on March 19 and 20, 1973. They had been working in the area and, when hearing about the lights, went to view them at a popular site to the east of Marfa, near the town of Alpine. The first night they saw lights rocking and looping as if "playing," they thought. The next night they drove without headlights in the same direction of the lights, hoping to sneak up on them. They observed three lights this time; one seemed to be aware of their presence, almost daring them to chase it! It moved low to the ground, darting

Marfa lights, photographed by James Crocker, September 1986.

around bushes, or hovered in the middle of the road, coming to within 200 feet of them. It resembled an ordinary household light bulb but was larger, about half the size of a basketball. After a short time the light moved to the east to join the others, and they all disappeared.

Some have explained Marfa's lights as reflections of stars and planets bent through layers of air of different temperatures or of headlights from cars passing along nearby Highway 67. While these explanations may account for some sightings, reports like the one above seem to challenge any reasoning. And there is no doubt that the Marfa lights have been around for quite a while: folklorists have collected stories that go back to the early white settlers and to the Apaches before them.

Sources:

Devereux, Paul, *Earth Lights Revelation: UFOs and Mystery Lightform Phenomena: The Earth's Secret Energy Force,* London: Blandford Press, 1989.
Stacy, Dennis, *The Marfa Lights: A Viewer's Guide,* San Antonio, Texas: Seale and Stacy, 1989.

Strange Showers: Everything but Cats and Dogs

- FALLS FROM THE SKY
- STRANGE RAIN
- ICE FALLS
- STAR JELLY

Strange Showers: Everything but Cats and Dogs

FALLS FROM THE SKY

For as long as human beings have been keeping records, things—both inorganic (composed of matter other than plant or animal) and organic (coming from living plant or animal matter)—have been falling out of the sky. While such falls usually take place in the middle of fierce storms, some do occur out of a clear sky.

Theories

For many centuries those who doubted that matter could fall from the sky believed the reports were mistakes, made by witnesses who failed to understand the natural process of *spontaneous generation*. Once widely believed, the concept of spontaneous generation held that living things could spring from nonliving material; thus, when rain hit the ground it could give rise—out of the mud, slime, and dust—to all sorts of living matter. This view eventually gave way to more realistic notions concerning the appearance of animals, vegetation, man-made objects, and other nonliving materials; doubters believed that these things were on the ground all along, unseen until rain drove or washed them into view.

Still, even more sophisticated theories failed to explain the many reliable accounts (some by scientists) of things falling from the sky. Thus other explanations arose, one suggesting that waterspouts (funnel- or tube-shaped columns of rotating, cloud-filled wind, usually with spray torn up from the surface of an ocean or lake), tornados, and whirlwinds (small rotating windstorms) pick up materials and drop

Waterspouts, from *L'Atmosphère* by Camille Flammarion, Paris, 1888.

them off somewhere else. While it is true that strong winds do rip objects from the ground and leave them elsewhere, this does not explain the odd *selectivity* seen in strange falls from the sky. Violent storms drop everything they pick up, but most falls drop only *one* thing or *one type* of thing. Often, too, the amount of material that descends is so great that its disappearance from another place would surely be noticed! Some falls go on for hours, with the given material falling in a steady stream over a very large area. While most falls of matter from the sky consist of dust or ash and can be easily explained, the cases that follow have scientists baffled.

Fire in the Sky

On October 18, 1867, residents of Thames Ditton, Surrey, England, were startled by a "shower of fire" in the evening sky. The light it cast was "brilliant" for the ten minutes that it burned. As reported in *Symons's Monthly Meteorological Magazine,* villagers found an explanation the next morning when "the waterbutts and puddles in the upper part of the village were thickly covered with a deposit of sulphur."

The fall of "a combustible substance of a yellowish color" also thought to be sulphur or a sulphur compound was reported in the village of Kourianof, Russia, some years earlier, in 1832. According to the *American Journal of Science,* villagers there first mistook the cottony material, which covered nearly seven hundred square feet of fields in a two-inch layer, for odd-colored snow. Samples, however, were quick to burst into flame and gave off a strange smell. This convinced witnesses that here was no ordinary snowfall! Because sulphur deposits are very rarely found on the earth's surface, scientists are doubtful that winds could pick up much and leave it somewhere else. Still, they can think of no other explanation for these strange fires from the sky.

Stones

In a March 1912 issue of the well-respected British scientific journal *Nature,* reporter G. E. Bullen described a strange happening that took place at Colney Heath, near St. Albans, Hertfordshire, England. During the stormy afternoon of March 4, witness H. L. G. Andrews watched a large stone hit the ground near him with such impact that it sunk down three feet, and a huge clap of thunder filled the air. The irregularly shaped rock weighed nearly six pounds and measured about 7 x 6 inches, and its surface was deeply pitted. Magnification showed that the rock was crystal and studded with rounded granules of nickel-containing iron. When Dr. George T. Prior of the British Museum of Natural History studied the stone, he concluded that it was "not of meteoritic origin."

Fish had fallen out of a clear sky in great numbers!

There are many more accounts of nonmeteoritic (not from meteorites) rocks and stones falling from the sky. Huge numbers of small black stones, for example, fell on Birmingham, England, in August 1858 and again at Wolverhampton, England, in June 1860. Both incidents took place during violent storms.

In late 1921 and early 1922, stones and rocks fell repeatedly on buildings and people in Chico, California. Events there began in November, when J. W. Charge, the owner of a grain warehouse along the Southern Pacific railroad tracks in Chico, complained to city marshal J. A. Peck that an unseen person was throwing rocks at his building daily. At first thinking the rock-throwing was a harmless prank, the marshal finally took action on March 8, when stones and rocks ranging in size from peas to baseballs battered the warehouse all day. He and his men searched the surrounding area but failed to locate a guilty party.

Whether somehow related to the falling stones or not, more than four decades earlier *fish* had fallen out of a clear sky in great numbers and landed on a roof and surrounding streets there!

Other rock falls have been recorded around the country. In 1943 Oakland, California, was the site of a stone-fall case very much like the one that took place in Chico. And showers of stones, described as warm to the touch, rained down on the pavement outside the office of the *Charleston News and Courier,* a South Carolina newspaper, at three separate times—2:30 and 7:30 A.M. and 1:30 P.M.—on September 4, 1886. According to witnesses, the stones fell straight down from the sky and only in a 75-square-foot area. Another stone-falling incident, which took place in a housing development outside Tucson, Arizona, lasted for four months, between September and December 1983. It was the subject of careful study by both police and D. Scott Rogo, an expert in the supernatural.

The coins seemed to come out of the sky; they would hear a tinkling sound on the sidewalk and, looking down, find a coin.

Pennies from Heaven

On the morning of May 28, 1982, a young girl walking through the yard of St. Elisabeth's Church in Redding, a small town near Manchester, England, saw a 50-pence coin fall "from nowhere." As the day went on other children found a number of coins at the same spot. Finally, the owner of a local candy store became concerned that the children were stealing from the church's poor box and told the Reverend Graham Marshall about the sudden rush to buy sweets. No money was missing, and the children all swore—when questioned by the clergyman—that the coins seemed to come out of the sky; they would hear a tinkling sound on the sidewalk and, looking down, find a coin.

Money has also fallen from the sky in other locations. One December day in 1968, shoppers in the English town of Ramsgate, Kent, heard pennies bouncing off the pavement. "Between 40 and 50 of them came down in short scattered bursts for about 15 minutes," one witness, Jean Clements, told the *London Daily Mirror.* "You could not see them falling—all you heard was the sound of them hitting the ground." The coins hit the ground hard enough to be dented. There were "no tall buildings nearby," Clements added, "and no one heard a plane go overhead."

Other money falls were reported in Meshehera, Russia, in the summer of 1940 (during a storm); in Bristol, England, in September 1956; in Bourges, France, April 15, 1957 (the "thousands" of 1,000-franc notes that fell were never claimed); and Limburg, West Germany, in January 1976 (2,000 marks, seen falling by two clergymen).

In 1969 a wheel sailed out of the sky and onto a California woman's car, leaving a one-foot dent.

Thunderstones

Some of the most fantastic claims are those concerning falls of human-made objects. Thunderstones (shaped stones such as axe heads), for example, were once the subject of worldwide folklore. Thunderstones came down right after a spectacular roar of thunder and bolt of lightning. In the modern world thunderstones have been replaced by other objects, like coins, that have been reported to drop out of the atmosphere.

Most of these human-made objects don't fall in clusters but by themselves. On April 17, 1969, the *New York Times* reported the bizarre experience of a California woman, Ruth Stevens, who was driving in Palm Springs when a wheel sailed out of the sky and onto her car's hood, where it left a one-foot dent. No local airport received a report of a missing wheel from any pilot.

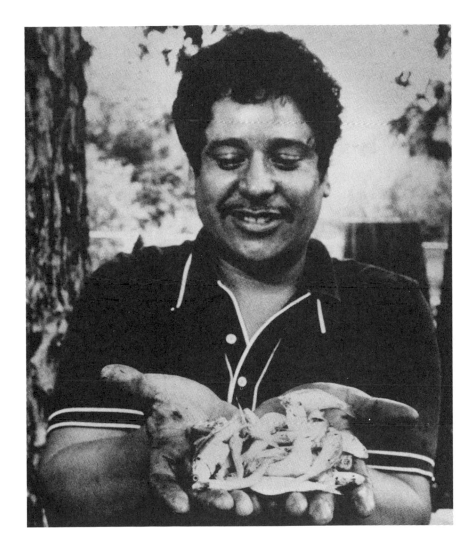

Louis Castoreno with some of the fish that fell in his yard at Fort Worth, Texas, in 1985.

Fish Stories

Late on the morning of February 9, 1859, residents of Mountain Ash, Glamorganshire, Wales, witnessed a strange event. As a heavy rain fell and a strong wind blew, a sea of fish fell out of the sky.

One observer, John Lewis, described his experience for the *Annual Register*: while getting ready to saw a piece of lumber, he "was startled by something falling all over me—down my neck, on my head, on my back. On putting my hand down my neck I was surprised to find they were little fish. By this time I saw the whole ground covered with them. I took off my hat, the brim of which was full of them. They were jumping all about." Covering a nearby shed and surrounding plants, the fish filled an area measuring "about 80 yards by 12." Witnesses gathered a bucketful of the fish, releasing them into a rain pool, where they happily swam about.

Between 7 and 8 A.M. on October 23, 1947, wildlife expert A. D. Bajkov and residents of Marksville, Louisiana, witnessed the fall of many thousands of fish, which landed—cold, and even frozen in some cases—on a strip of ground 75 feet wide and 1,000 feet long. Weather conditions were foggy but calm, which makes this case unusual, for most falls take place during storms.

While the Marksville fish were, according to biologist Bajkov, identical to those found in local waters, Science writer J. Hedgepath witnessed a brief fish fall in Guam in 1936 where one of the specimens was identified as a tench, common only to the fresh waters of Europe!

At Nokulhatty Factory, India, on February 19, 1830, a great number of fish dropped from the sky. Writing in the American Journal of Science, M. Prinsep reported: "The fish were all dead; most of them were large; some were fresh, others rotted and mutilated. They were seen at first in the sky like a flock of birds descending rapidly to the ground. There was rain drizzling at the time, but no storm." Some of the fish had no heads.

Toads

On September 23, 1973, *tens of thousands* of toads fell on Brignoles, France, during what was described as a "freak storm." They were all young toads. In September 1922 young toads fell for two days on another French village, Chalon-sur-Saone.

Especially fascinated by accounts of living material falling from the sky, investigator of odd physical events Charles Fort collected some 294 of them before his death in 1932. Many more have been reported since. The following demonstrates just *some* of the living creatures involved in falls. Still, with the exception of the turtle incident, falls of these items have occurred not once, but again and again!

Alligators

December 1877: At a Silvertown Township, South Carolina, turpentine farm—set on high, sandy ground about six miles north of the Savannah River—Dr. J. L. Smith noticed something fall and begin to crawl toward the tent in which he sat. "On examining the object he found it to be an alligator," reported the *New York Times* (December 26). "In the course of a few moments, a second one made its appearance. This so excited the curiosity of the Doctor that he looked around

On September 23, 1973, tens of thousands of toads fell on Brignoles, France, during what was described as a "freak storm."

Charles Hoy Fort is recognized for his study of odd physical events. Until Fort's undertaking, beginning in the late nineteenth century, no one had collected and organized the numerous reports of physical anomalies.

Charles H. Fort

(1874-1932)

A struggling journalist, Fort spent much of his time in libraries, where he read widely and took notes mostly on oddities of nature or sightings of objects that pointed to the presence of otherworldly intelligences. He wrote four books on anomalies: *Book of the Damned* (1919), *New Lands* (1923), *Lo!* (1931), and *Wild Talents* (1932). In all of them, Fort offered wacky "theories" to explain strange events, trying to show that his ideas were really no weirder than those scientists used to try to "explain away" happenings they did not understand.

So important were Fort's books, in fact, that the odd events he described came to be known as "Fortean phenomena." A Fortean Society, formed in 1931 in the United States to collect and record "quite extraordinary occurrences," was later replaced by the International Fortean Organization.

to see if he could discover any more, and found six others within a space of 200 yards. The animals were all quite lively, and about 12 inches in length."

Blood and flesh

July 1869: As mourners prepared for a funeral at a farm near Los Angeles, blood and meat rained out of a clear sky for three

BY CHARLES FORT

minutes. It covered two acres of a cornfield. When examined, the blood was found to be mixed with what looked like hairs from animal fur. The flesh ranged in size from small particles to six- and eight-inch strips and included what observers believed were pieces of kidney, liver, and heart. One witness brought samples to the *Los Angeles News,* the editor of which later wrote in the August 3 issue, "That the meat fell, we cannot doubt. Even the parsons in the neighborhood are willing to vouch for that." (Also see entry: **Strange Rain**.)

Grain

March 24, 1840: During a thunderstorm, a shower of grain covered Rajket, India, and a large area of the surrounding countryside. According to the *American Journal of Science,* a British captain named Aston collected samples and sent them to England. The man related that Indian witnesses were upset not only by the frightening sight of grain falling from the sky, but by the fact that the grain was not even grown in their country—it was of a type entirely unknown to them!

Green slime

September 5 and 6, 1978: An "unexplained green slime," according to the *Journal of Meteorology,* splattered an area of Washington, D.C., injuring animals and plants and soiling automobiles. The roof of a 12-story building was coated with the substance, suggesting it had fallen from great height. Officials did not even try to explain it.

Hazelnuts

May 9, 1867: Hazelnuts "fell in great quantities and with great force" over a small area of Dublin, Ireland. They came down so hard, in fact, "that even the police, protected by unusually strong head covering, were obliged to take shelter," wrote a reporter in *Symons's Monthly Meteorological Magazine.*

Leaves

April 7 and 11, 1894: On two clear, still days, leaves fell for half an hour on two French villages, first at Clairvaux, then at Pontcarre. Though it was spring and trees were just starting to bud, the fully grown leaves were dried and dead—autumn leaves.

Lizards

December 1857: During a rain shower, lizards fell on streets and sidewalks in Montreal, Quebec.

Mussels

August 9, 1892: A rapidly moving yellow cloud suddenly released crushing rains, as well as hundreds of mussels onto the streets of Paderborn, Germany.

Salamanders

June 1911: Arlene Meyer recalled that once, when she was a girl, she was walking along the banks of the Sandy River near Boring, Oregon, when she got caught in a sudden rainstorm. Feeling large objects hit her on the head and shoulders, she looked around her and was shocked to see hundreds of salamanders "falling from the sky, literally covering the ground and wriggling and crawling all over."

Seeds

Summer 1897: Shortly before sunset a large number of small, dark-red clouds filled the sky over Macerata, Italy, and an hour later a powerful storm erupted. Immediately, "the air became filled with myriads of small seeds," R. Hedger Wallace reported in *Notes and Queries,* which "fell over town and country, covering the ground to a depth of about half an inch." The following day scientists at Macerata began looking for explanations. Prof. Cardinali, a respected Italian naturalist, identified the seeds as those of a plant commonly called the Judas Tree, found only in central Africa or the Antilles (in the West Indies, bordering the Caribbean Sea)! Even more incredible, a great number of the seeds had already begun growing!

Straw

August 1963: Straw in huge amounts began falling from clouds over Dartford, Kent, England, and continued for an hour. "I looked up," one witness said, "and the sky was full of it." The fall stopped as suddenly as it had begun. "We are mystified," a government weather expert told the Associated Press. A police officer commented on the straw, insisting, "There was far too much of it for it to have been dropped from an airplane."

Turtle

May 11, 1894: A gopher turtle, six by eight inches, fell during a hailstorm at Bovina, Mississippi.

Worms

July 25, 1872: The magazine *Nature* reported that at 9:15 on a hot evening with a clear sky, "a small cloud appeared on the horizon, and a quarter of an hour afterwards rain began to fall, when, to the horror of everybody, it was found to consist of black worms the size of an ordinary fly. All the streets were strewn with these curious animals."

More Theories

If waterspouts and whirlwinds are not responsible for the fall of plants and animals from the sky—or if specimens are not actually earthbound objects mistaken by witnesses—what other theories could explain such a relatively widespread oddity? Few have tried to supply answers to so puzzling a question, and those without much success.

Fort, for example, had his own favorite theory, which he cheerfully admitted was ridiculous and unscientific. He suggested that giant land masses floated above the earth, and tornadoes, hurricanes, and cyclones carried all sorts of items upward and dumped them on these lands. He also described a "Super-Sargasso Sea," an ocean in the atmosphere that contained castoffs from other times and even other worlds—a crazy mix of spaceship wrecks, dinosaurs and fossils, "horses and barns and elephants and flies"—just about anything one could imagine. And just as fierce earth storms could dump things onto these heavenly land masses or into its sea, they could also force the fall of all sorts of organic and inorganic matter back to earth.

Other explanations include John Philip Bessor's theory involving space animals. Because of the "many falls of flesh and blood from the sky," Bessor believed that UFOs were really meat-eating atmospheric life forms, or space animals. According to him, this idea also explained the mysterious disappearances of people! Morris K. Jessup, a UFO writer, felt that the fall of living things from the sky occurred when space beings—who collected and raised earth creatures in special hydroponic tanks in outer space—needed to empty the tanks for cleaning or to rid them of extra specimens.

Other writers have tried to make sense of falls from the sky by using the idea of teleportation—the instant movement of an object from one place to another. But as teleportation is unproven, trying to explain one mysterious occurrence with another is rarely convincing! And just like other theories, teleportation doesn't explain the baffling selectivity seen in falls or the huge amounts of things that fall in many cases.

Damon Knight, a biographer of Charles Fort, added scientific research to Fort's playful theory about falls and came up with an interesting idea of his own. He matched weather records with the reports of falls and unusual space and sky happenings in Fort's books and found strong connections: the years 1877 through 1892, for example, had both an unusually high number of odd reports and some extraordinary weather. He also studied physics and astronomy and the effect of solar system activities on everything from weather patterns to human behavior. He came to the conclusion that "under certain conditions of gravidic [gravity] and electromagnetic strain in the solar system, channels open through which material objects can reach the Earth from parts unknown, or can be transferred from one part of the Earth's surface to another." Referring to the puzzling selectivity seen in most falls, Knight added, "All living things have electric charges, and it is possible to imagine that an electromagnetic field would discriminate between them."

Sources:

Fort, Charles, *The Books of Charles Fort,* New York: Henry Holt and Company, 1941.

Knight, Damon, *Charles Fort: Prophet of the Unexplained,* Garden City, New York: Doubleday and Company, 1970.

Michell, John, and Robert J. M. Rickard, *Phenomena: A Book of Wonders,* New York: Pantheon Books, 1977.

Rogo, D. Scott, *On the Track of the Poltergeist,* Englewood Cliffs, New Jersey: Prentice-Hall, 1986.

STRANGE RAIN

It is hard to imagine that something as commonplace as rain can be full of mystery. Yet some rainfalls have been so odd that they have frightened witnesses and puzzled scientists.

One kind of strange report, though rare, involves the falling of water from clear skies. In October 1886 in Charlotte, North Carolina, the local newspaper ran a story about a patch of land between two trees that had received a rain shower every afternoon, for three weeks, whether the sky was cloudy or clear.

A worker for the U.S. Signal Corps arrived to check out the account and was amazed to find the report true. Writing in the October issue of *Monthly Weather Review,* he noted—on the first day—"rain drops at 4:47 P.M. and 4:55 P.M. while the sun was shining brightly." The second afternoon he returned. Between 4:05 and 4:25, "a light shower of rain fell from a cloudless sky." Sometimes the rain fell "over an area of half an acre," but it always seemed "to center at these two trees, and when lightest [fell] there only." Not long after the signal corps visit, the odd rain stopped as mysteriously as it had begun. But a similar happening occurred that October in Aiken, South Carolina. Rain fell from morning till late at night on two graves in the town cemetery—and nowhere else—with not a cloud in the sky. Witnesses there numbered in the hundreds.

Science usually explains such rain from clear skies as water blown from another, cloudier, place. This doesn't explain why the rain falls on the exact same spot again and again, however! An even more common type of strange rainfall involves colored rain. There are accounts of blood-red rains that have completely baffled scientists!

Blood-Red Rains

In July 1841, for example, slaves in a field in Wilson County, Tennessee, reported a small red cloud in an otherwise clear sky. From the cloud fell a shower of blood droplets and small pieces of muscle and fat. The local doctor, W. P. Sayle, took samples from the site and sent them to a professor of chemistry at the University of Nashville for study. Physician G. W. Bassett of Virginia reported a similar event in the spring of 1850. On Good Friday, a small cloud passed over a friend of the doctor's, who was with several servants. Pieces of flesh and liver rained to the ground. The next morning Bassett collected

Rain fell from morning till late at night on two graves in the town cemetery—and nowhere else—with not a cloud in the sky.

15 to 20 pieces and sent some to a fellow doctor in an effort to find out what kind of flesh it was. The rest he preserved in alcohol for future examination.

Other accounts of bloody rain appeared in the nineteenth century. On February 15, 1849, in Simpson County, North Carolina, fresh pieces of flesh, liver, brains, and blood fell out of a red cloud and splattered over an area 30 feet wide and 250 to 300 yards long. The *San Francisco Herald* of July 24, 1851, reported a two- to three-minute shower of blood and flesh, with pieces ranging in size from a pigeon's egg to a small orange, that fell on an army station at Benicia, California. And a bloody rain containing hairs and portions of organs fell on two acres of a cornfield outside Los Angeles in July 1869; it was witnessed by a funeral party, which included several members of the clergy.

On March 8, 1876, "flakes of meat" fell from the sky and landed on a Bath County, Kentucky, field, and one brave witness tasted a "perfectly fresh" sample. It reminded him, he told *Scientific American,* of "mutton or venison." The case sparked much attention—and two weak explanations. One stated that the matter befouling the field was a form of algae, which had been on the ground all along but sprouted when the rain soaked it; in fact, the sky was clear during the fall. The second rationale was that the material was buzzard vomit, even though it fell in flakes of one to four inches square and covered ground, trees, and fences on a strip of land 100 yards long and 50 yards wide.

Locales outside the United States have experienced blood-red rains as well. On October 16 and 17, 1846, French witnesses were terrified by a blood shower. After red rain fell in Messignadi, Calabria, Italy, weather experts identified it as bird blood. And in 1888 two red rains fell on the Mediterranean region, 12 days apart. Samples that were burned emitted the strong "odor of animal matter," according to the French scientific journal *L'Astronomie.* In the only twentieth-century account on record, meat and blood showered two small towns between São Paulo and Rio de Janeiro, Brazil, on August 30, 1968. Said a police officer who witnessed the scene, "The sky at the time was quite clear. No aircraft had been seen just prior to, during, or after the event, nor were there any birds in the sky."

It is not known whether bloody rains happen less often in our time or whether witnesses are too reluctant now, in our more scientific age, to report such unbelievable sights.

Sources:

Constance, Arthur, *The Inexplicable Sky,* New York: The Citadel Press, 1957.
Corliss, William R., ed., *Handbook of Unusual Natural Phenomena,* Glen Arm, Maryland: The Sourcebook Project, 1977.
Corliss, William R., ed., *Tornados, Dark Days, Anomalous Precipitation, and Related Weather Phenomena: A Catalog of Geophysical Anomalies,* Glen Arm, Maryland: The Sourcebook Project, 1983.

ICE FALLS

On the evening of September 2, 1958, in his home in Madison Township, New Jersey, Dominick Bacigalupo stood up from a kitchen chair and took a step or two just before his roof caved in. Unhurt but shaken, the man looked about him and realized what had happened: a 70-pound chunk of ice had crashed through the top of his house and had fallen in three big pieces in the cooking area.

It was not storming that night. Bacigalupo's 14-year-old son, Richard, had noticed two airliners flying by just before the bizarre fall, but airport officials denied that the planes were carrying ice. Weather experts at nearby Rutgers University said conditions in the atmosphere could not have created ice chunks of such size and weight. So where did the ice come from?

Ice chunks dropping from the sky is one of meteorology's most frequent and puzzling mysteries. Usually weather experts explain such falls as the result of ice buildup on planes. But this explanation fails to satisfy for many reasons. One is that the electrical heating systems on most modern aircraft prevent ice buildup on wings or other surfaces. Also, according to the Federal Aviation Agency, even older planes—without such heating systems—rarely build up large amounts of ice because of the way they are constructed and because of their swift flying speed. More important, some reported ice chunks have been so huge and heavy that any aircraft carrying them, even for a short time, would be in serious danger of crashing!

Early Reports of Ice Falls

In fact, large blocks of ice have been falling from the sky since long before the invention of the airplane. Late in the eighteenth century, for example, a chunk "as big as an elephant" reportedly fell on Seringapatam, India, and took three days to melt. As unbelievable as this may sound, accounts of enormous falling ice blocks have been recorded frequently.

An 1849 issue of the *Edinburgh New Philosophical Journal* noted that on one August evening that year, a large mass of ice fell on the Balvullich farm, on the estate of Orda in Scotland, measuring nearly 20 feet around and 20 feet thick! A resident farmer named Moffat reported a huge clap of thunder as the irregularly shaped chunk fell near his house. Examination of the block demonstrated that it "had a beautiful crystalline appearance being nearly all quite transparent ... except [for] a small portion of it which consisted of hailstones of uncommon size,

Late in the eighteenth century, for example, a chunk "as big as an elephant" reportedly fell on Seringapatam, India, and took three days to melt.

Beryl Voyle displays a small portion of a gigantic block of ice that fell in a nearby field in Wales, 1986.

fixed together. It was principally composed of small squares, diamond-shaped, of from 1 to 3 inches in size, all firmly congealed together." While observers could not figure out how to weigh the huge mass of ice, all agreed that Moffat and his household had been very lucky, narrowly avoiding being crushed to death! Oddly enough, no hail or snow was reported anywhere else in the district that day.

On December 26, 1950, another Scottish man, driving near Dumbarton, watched as a mass of ice fell from the sky—nearly hitting him—and crashed to the road. When police arrived at the scene and gathered the pieces, they weighed 112 pounds. This was just one of many ice falls that took place in Great Britain over a two-month period between November 1950 and January 1951. In Kempten, West Germany, also in 1951, tragedy struck when a block of ice six feet long and six inches around fell on a carpenter working on a roof and killed him. In February 1965 a 50-pound mass of ice smashed through the Phillips Petroleum Plant roof in Woods Cross, Utah.

It is interesting to note that the largest recorded hailstones are slightly over five inches around and weigh a little more than two pounds. Hailstones fall, of course, during storms, where winds and

updrafts keep them aloft before they hit the ground. Ice falls, on the other hand, most often come out of clear skies.

Scientists Investigate

One of the best-studied cases of falling ice was reported by British meteorologist R. F. Griffiths in 1973. On April 2 of that year, while waiting at a street crossing in Manchester, England, the scientist saw a large object strike the road near him and shatter into pieces. He picked up the largest chunk, weighing three and a half pounds, and rushed home to store it in his freezer. He later wrote that the sample proved puzzling, because "whilst it is clearly composed of cloud water, there is no conclusive evidence enabling one to decide precisely how it grew.... In some respects it is very much like a hailstone, in others it is not." A look at flight records for the area showed that no planes had passed overhead at the time.

While not sure of a connection, Griffiths reported that the fall took place nine minutes after another strange happening in the sky: a "single flash of lightning." Many others noticed it as well "because of its severity, and because there were no further flashes." The scientist also noted "unusual meteorological conditions" in England that day, including gales and heavy rains. Manchester itself had had snow in the morning, but skies were clear at the time the ice fell; sleet followed later that day. In a 1975 issue of *Meteorological Magazine*, Griffiths stated his determination that the lightning was triggered by a plane flying into a storm further to the east. About the ice sample, however, he could draw no conclusion.

Another well-studied ice fall that proved even more puzzling took place on a farm in Bernville, Pennsylvania, in 1957. Early on the evening of July 30, farmer Edwin Groff heard a "whooshing noise" and looked up to see a large, white, round object sailing out of the southern sky. After it crashed and shattered a few yards from him, a second, similar, object struck a flower bed close to where he and his wife stood. The first of these was a 50-pound ice cake; the second was half the size and weight of the first.

Witnesses immediately called Matthew Peacock, a meteorologist who lived in nearby Reading. The scientist asked a colleague, Malcolm J. Reider, to examine the fallen ice. It was cloudy and white, as if it had been frozen rapidly, and was filled with "sediment"—dust, fibers, algae. Put together like a "popcorn ball," the ice chunk appeared to be made up of many one-inch hailstones frozen together in a single mass. Yet hailstones do not contain such sediment.

Chemical analysis also showed that the ice mass lacked iron and nitrate, two elements always present when ordinary groundwater is rapidly frozen. Indeed, the specimen seemed to be related neither to ordinary precipitation nor to water swept up from the ground. Another scientist who examined the ice chunk, Paul Sutton, chief of the U.S. Weather Bureau station at Harrisburg, declared that the ice was "not formed by any natural process known to meteorology."

Theories

Charles Fort, one of the first persons to collect and study reports of such anomalies, made clear—with his many scientific journal accounts—that ice falls were a fairly common meteorological oddity. His own theory, made half-jokingly, was that "floating in the sky of this earth, there are fields of ice as extensive as those on the Arctic Ocean"; violent thunderstorms sometimes loosen pieces and they fall below (also see entry: **Falls from the Sky**).

William R. Corliss is a physicist who has written a number of books on rocketry and space travel. He is best known, however, for the Sourcebook Project, an enormous collection of articles and reports about physical and natural mysteries. More oriented to conventional science than the unorthodox Charles Fort—the first great investigator of physical anomalies, who died in 1932 (also see entry: Falls from the Sky)—Corliss researched mainly from scientific publications. Over the years he has collected thousands of accounts of unusual happenings, concentrating mostly on subjects like unusual weather and geophysical (earth-related) oddities—occurrences that, while important, are less likely to outrage scientists than the UFOs and monstrous creatures that delighted Fort.

Beginning with the publication of *Strange Phenomena* in 1974, Corliss began reprinting some of the articles he had uncovered. Nearly a decade later, he had published 16 volumes and well over 7,000 pages of source material on scientific anomalies. And this was only a small part of his collection! In 1982 Corliss stated: "The cataloging task is just beginning, for the anomalies in the world's scientific and semiscientific literature seem nearly infinite in number." Corliss also publishes a newsletter on current anomalies, *Science Frontiers*.

William R. Corliss

Other, more recent, theories center on UFOs; ufologist M. K. Jessup, for instance, offered this explanation of ice falls: "It seems most natural that a space contrivance [vehicle], if made of metal, and coming from cold space, would soon become coated with ice. That the ice should fall off, or be pushed off by de-icing mechanisms, or even melt off when the space ships are heated by friction with the air or become stationary in the sunshine, seems equally natural." But, in actuality, few ice fall reports include UFO sightings.

Scientists generally rely on two theories concerning ice falls. One suggests that the ice originates somewhere within the earth's atmosphere; strange-weather expert William R. Corliss, for example, feels that "some unappreciated mechanism in hailstorms permits the sudden aggregation [clustering] of many hailstones." The second theory—once ridiculed but now taken more seriously—suggests that ice chunks are true meteorites, coming from outer space. The only problem with this idea, according to critic Ronald J. Willis, is that "there is little indication of high speed entry into the atmosphere that we would expect from any meteorite, whatever its origin." Because fallen ice chunks vary so much in appearance and composition, perhaps more than one such theory is needed to explain them.

Sources:

Corliss, William R., ed., *Handbook of Unusual Natural Phenomena,* Glen Arm, Maryland: The Sourcebook Project, 1977.
Fort, Charles, *The Books of Charles Fort,* New York: Henry Holt and Company, 1941.
Hitching, Francis, *The Mysterious World: An Atlas of the Unexplained,* New York: Holt, Rinehart and Winston, 1978.
Lorenzen, Coral E., *The Shadow of the Unknown,* New York: Signet, 1970.

STAR JELLY

P*wdre ser,* or star jelly, is a strange phenomenon that has been around for centuries. The Welsh phrase *pwdre ser* means "rot from the stars," and it describes the glowing, jellylike material that witnesses observe on the ground after seeing strange lights or meteorlike objects move across the sky. In 1541 the poet Sir John Suckling wrote a poem about the oddity, as did writer John Dryden in 1679. While unable to explain pwdre ser, men of science—back then, at least—had no doubt that the oddity was real. Scientists today, however, are not so sure because of what they know about meteors and meteorites; they believe that no such matter could survive the fiery trip through the earth's atmosphere.

Still, consider this story, recorded in the *Transactions of the Swedish Academy of Sciences* in 1808: on May 16 of that year, on a warm, cloudless afternoon, the sun dimmed for an unknown reason. Slowly, from the west, a great number of dark brown balls, about the size of the top of a hat, rose and approached the sun. They became darker and for a brief time stopped moving and hung in the air; when they resumed motion, they picked up speed and traveled in a straight line until they disappeared in the east. During that time, some vanished and others fell down. The strange parade continued in the same manner for nearly two hours. There was no noise—banging, whistling, or buzzing—in the air. Tails nearly 20 feet in length trailed some of the faster balls; but these slowly faded away.

Some of the balls fell to earth not far from K. G. Wettermark, an official of the Swedish Academy. The man noted that as the objects fell they lost their dark color, disappeared briefly, then became visible again but in changing colors—like "those air-bubbles which children ... produce from soapsuds." Wettermark added, "When the spot, where such a ball had fallen, was immediately after examined, nothing was to be seen, but a ... film ... as thin and fine as a cobweb, which was still changing colors, but soon entirely dried up and vanished."

Another case of pwdre ser was reported in the *American Journal of Science* in 1819 by a Prof. Rufus Graves. Earlier that year, between 8 and 9 P.M. on August 13, a fireball "of a brilliant white light resembling burnished silver" fell slowly from the sky and onto the front yard of an Amherst, Massachusetts, man named Erastus Dewey. Two neighbor women also noticed the light. In the morning Dewey found a strange material 20 feet from his door (which Graves soon examined). Circular in form, eight inches across and one inch high, the tan blob had a fuzzy coating that, when removed, revealed a soft center that gave off an awful smell—so awful, in fact, that it made witnesses dizzy and sick to their stomachs! After awhile the blob turned a blood-red color and began to absorb moisture from the air. Some of the material was collected in a glass, where it began to change into a paste, taking on the color, look, and feel of regular household starch. Within two or three days, the material had disappeared from the glass altogether. All that remained was a dark-colored film that, when rubbed between the fingers, turned into a fine, odorless ash.

Reports of the British Association also printed an account of pwdre ser that occurred late on the evening of October 8, 1844, near Coblentz, Germany. Two men walking in a plowed field were startled to witness the fall of a glowing object that crashed to earth not 20 yards from them. Because it was too dark to investigate that night, they marked the spot and returned to it early the next morning.

> ## METEORS
>
> A *meteor* is one of the small pieces of matter in the solar system that can be seen only when it falls into the earth's atmosphere, where friction may give it a temporary glow (incandescence). It is sometimes called a "falling" or "shooting" star. A *meteorite* is a meteor that reaches the surface of the earth without being completely destroyed by the heat of friction.

One night a bright light fell over the neighborhood. The next morning Mrs. Christian found three purple blobs on the front lawn.

Expecting to find the remains of a meteorite, the two, instead, found a gray, jellylike mass that shook when they poked it with a stick. They did not try to preserve it.

Science Looks for Other Explanations

By the 1860s, as scientists came to learn more about meteors and meteorites, they began to doubt pwdre ser's connection with the stars and looked, instead, for more earthly explanations. Some scientists thought the material might be bird vomit; plant specialists believed it might be *nostoc,* a blue-green algae. Edward Hitchcock, an Amherst College chemistry professor, was sure, in fact, that the material in the Amherst case was a "species of gelatinous fungus, which I had sometimes met with on rotten wood in damp places." The professor believed it was an "entire mistake" that had caused observers to connect it with the falling object. Still, not too long ago scientists doubted that meteorites were real. They thought it an "entire mistake" for observers to connect stones and rocks with the glowing "falling stars" that they had seen go down at the exact same spots. Now, of course, no one doubts meteorite witnesses.

Reports of pwdre ser in our time are rare. The most famous recent case took place in 1979 in Frisco, Texas, outside Dallas. On the night of August 10, a bright light fell over the neighborhood where Martin and Sybil Christian lived. The next morning Mrs. Christian found three purple blobs on her front lawn. One of the blobs dissolved; the other two were frozen and sent away to be studied. They were identified as waste from a nearby battery factory, but the factory strongly disagreed with this finding and refused to accept blame. Chemical reports, in fact, demonstrated that the purple blobs were very different from the factory's waste, and the two materials did not even appear similar.

Sources:

Burke, John G., *Cosmic Debris: Meteorites in History,* Berkeley, California: University of California Press, 1986.
Corliss, William R., ed., *Handbook of Unusual Natural Phenomena,* Glen Arm, Maryland: The Sourcebook Project, 1977.

More Weird Weather

- **STRANGE CLOUDS**
- **BALL LIGHTNING**
- **SKYQUAKES**
- **TUNGUSKA EVENT**

More Weird Weather

STRANGE CLOUDS

One pleasant summer morning in 1975, science teacher Tom D'Ercole of Oyster Bay, New York, was in his driveway and about to enter his car when he glanced up at the sky. There, hovering above the roof of his house, he saw a small dark cloud different from the other few clouds that were floating by much higher in the sky. D'Ercole reported that the "cloud" seemed to move and grow as he watched it. At first the size and shape of a basketball, it floated back and forth across the peak of the roof, then became egg-shaped, then became "multicurved, dark, [and] vaporous." D'Ercole observed that "it finally measured about six feet in height and one and one-half feet in width."

Stunned and unable to think of an explanation, he continued to watch in disbelief as things got even stranger. The cloud seemed to inhale, pucker its "lips," and direct a stream of water toward him and the car, soaking both! After a minute the spray stopped, and the cloud disappeared instantly. After changing his clothes, D'Ercole took his wet shirt to the junior high school where he taught to run chemical tests on it. The wet substance was indeed merely water.

This event seems like nature's idea of a practical joke. Yet clouds *are* capable of strange appearances and behaviors, which science can sometimes explain—and sometimes cannot. In his study *Tornados, Dark Days, Anomalous Precipitation, and Related Weather Phenomena* (1983), physicist William R. Corliss looked at scientific reports of cloud arches, glowing clouds, rumbling clouds, clouds with holes in them, and more. Though the reports are puzzling, meteorologists have ideas about the causes behind most cloud phenomena. But other reports have revealed clouds so odd that no explanations seem possible!

The cloud seemed to inhale, pucker its "lips," and direct a stream of water toward him and the car, soaking both!

Nature's idea of a practical joke?

Runaway Clouds

One interesting case reported in several scientific journals involved a small, slow-moving, perfectly round white cloud that suddenly appeared in an otherwise clear sky northwest of Agen, France, in the late morning of September 5, 1814. After a few minutes, it stopped and remained still for a time before suddenly speeding southward, all the while spinning and making ear-shattering rumbling noises. It then exploded and rained down a variety of stones, some quite large! The cloud stopped its movement then and slowly faded away. Incredibly, similar events have been recorded in Sienna, Italy, in 1794; Chassigny, France, in 1815; Noblesville, Indiana, in 1823; and elsewhere.

Cigars in Clouds

Just before the wave of UFO sightings that took place in France in the autumn of 1954, several witnesses—including a businessman, two police officers, and an army engineer—observed an extraordinary sight over the town of Vernon. Watching from his driveway at 1 A.M. on August 23, businessman Bernard Miserey saw a huge vertical cigar, 300 feet long, hovering above the north bank of the Seine River 1,000 feet away. According to his testimony, "a horizontal disc" suddenly dropped from the bottom of the cigar, swaying and diving in the air. As it approached him it became "very luminous" before vanishing in the southwest. Over the next 45 minutes other, similar discs dropped out of the cigar. By this time the mother craft had lost its glow and disappeared into the darkness.

Clouds—or more specifically, "cloud cigars"—were described in a number of UFO sightings that occurred from the late 1940s into the 1960s. Usually these objects were seen with smaller, disc-shaped structures; thus the cloud cigars came to be viewed as the "motherships."

Although no clouds were mentioned in that sighting, it set the scene for an even more spectacular event. This one, with hundreds of witnesses, took place three weeks later on September 14, in the southwest of France along the Atlantic coast. At 5 P.M., while working with his men in a field, a wealthy farmer who lived near Saint-Prouant saw a "regular shape something like a cigar or a carrot" drop quickly out of a thick layer of clouds. The object was horizontal, luminous, and rigid, and it did not move with the clouds above it. It looked, farmer Georges Fortin said, like a "gigantic machine surrounded by mists." When the object completed its descent, it moved into a vertical position and became still.

By now citizens of half a dozen local villages, as well as farmers living in the region, were watching the sky in awe. White smoke or vapor began to pour out of the bottom of the object, shooting straight down before slowing and rising to circle the cigar in spirals. When the wind had blown all the smoke away, its source was revealed: a small metal disc that shone like a mirror and reflected light from the larger object. The disc darted about the area, sometimes moving with great speed, sometimes stopping suddenly, before finally streaking toward the cigar and disappearing into its lower part.

Fortin reported that about a minute later, the "carrot" leaned back into "its original position, point forward." It sped ahead "and disappeared into the clouds in the distance." The whole event had lasted "about half an hour." Other witnesses up and down the valley repeated this account. Weather experts reported that no tornado or other unusual meteorological activity had taken place at the time of the sighting.

A 300-foot-long, dull-gray, cigar-shaped machine came out of a cloud during a rainstorm over Cressy, Tasmania, Australia, on October 4, 1960. Among those who saw it was Lionel Browning, an Anglican

Spectacular cloud formation over Mount Shasta, California.

minister. As he and his wife watched the object, which they guessed was 300 feet above the ground, five or six domed discs measuring about 30 feet across shot out of the clouds just above and behind the cigar. They headed toward it "like flat stones skipping along water"— exactly how pilot Kenneth Arnold described the motion of the discs he saw over Mount Rainier, Washington, on June 24, 1947, in a sighting that would launch the UFO age (also see entry: **Unidentified Flying Objects**). According to Browning, the ship seemed to sail on "for some seconds unaware that it had shed its protection. Possibly when this was discovered, the saucers were called to the mother ship. The objects then moved back into the cover of the rain storm."

UFO-like Clouds

The sighting of a very odd cloud was recorded in the log of the ship *Lady of the Lake* on March 22, 1870, as it was sailing off the coast of West Africa, south of Cape Verde. Called by his men to the rear of the ship to witness the strange sight, Captain F. W. Banner observed a round cloud with five rays or arms extending from its center. Light gray in color, it had a tail similar to that of a comet. Moving along much lower than the other clouds in the sky, it was visible for nearly a half hour before darkness fell.

In another strange sighting nearly a century later, a pair of clouds resembling "puffy-like daubs of cotton" passed over Sunset, Utah, late on the afternoon of October 14, 1961. The clouds were linked by a cord of long, stringy material. Behind them were two smooth, metallic, disc-shaped structures. All four objects disappeared over the horizon. Ronald Miskin, an investigator for the Aerial Phenomena Research Organization, interviewed witnesses the next day. One was Sunset's mayor, who was pointing skyward and describing the objects' path when a "puffy" white object again flew overhead and was quickly joined by another. They streaked across the sky in the same direction taken by the objects seen the day before.

Aliens in the Clouds

Late one afternoon in the spring of 1965, a tourist looking out the window of a clifftop house along the seashore in Sydney, Australia, noticed a beautiful pink cloud that did not move. An hour later, when she looked again, the cloud was moving in her direction. It was soon below her eye level, which allowed her to look down on it, and to her amazement she saw a round, white object inside. Vents along the object's side gave off gray steam, which, as it covered all but the top, turned pink. The object appeared to be making its own cloud!

As if this were not fantastic enough, an engine sound came from the still-lowering object. A glowing ladder dropped from its underside, and a humanlike figure climbed down to a lower rung. There it sat and directed a searchlight toward the sea below. Some distance out on the water, a pink flare shot into the air. The ladder pulled in at once, and the object shot off in the direction of the flare. The witness then noticed a long (though not clearly visible) shape in the water from where the flare had come. Both the UFO and the underwater shape vanished in a "vivid pink flash."

On the afternoon of January 7, 1970, two Finnish skiers came upon a mysterious glowing red "cloud." When it got within 50 feet of them, the cloud seemed to dissipate somewhat, and they saw that a smoke-blowing domed disc was at its center. The object hovered near them and in the light it cast, they could see a three-foot-tall humanoid with a pale, waxy face, no eyes, and a hooklike nose standing on the ground just beneath it. After about 20 seconds, the red fog suddenly reappeared, and by the time it cleared, both the object and the being were gone.

"Saucer Clouds" seen over Marseilles, France, in 1955.

Although there have been other cases in which "clouds" or "fogs" have played a role in meetings with alien beings, they are not common in such reports.

Mysterious Planes and Vanishing Clouds

A drought that began in 1973 and continued for well over a decade was the starting point for a number of strange "sightings" that were reported by Spanish farmers in three southern provinces in the early 1980s. With their thirsty farmland all but useless, they began charging that the lack of rainfall was not an unfortunate condition of nature but an evil conspiracy planned by the big tomato growers. The small farmers were certain that these growers did not want rain to fall on their humble crops—and that to make certain of this, the tomato growers had hired pilots to destroy rain clouds.

More Weird Weather

Despite the fact that no technology exists to break up rain clouds, the farmers stood by their story. Assurances by scientists, legal officers, and flying experts (who swore that small planes could not fly into storm clouds without serious risk of crashing) meant nothing. The farmers insisted that on many occasions they had seen the appearance of a thundercloud on the western horizon, followed within minutes by the approach of an unmarked aircraft. The aircraft would fly into the cloud, leave off chemicals, and reduce it to wisps.

A drought in southwestern France in 1986 brought similar complaints. This time the villain was said to be a corporation developing anti-hail technology. It did no good for experts to report that nothing could be done to stop hail. Fortunately, heavy rains that summer put an end to the affair!

To social scientists, these cases demonstrate how stressful conditions can stir people's fears and imaginations. What is odd, though, is that even individuals not directly affected by the drought said they had seen the planes in action. One of them, agriculture ministry engineer Francisco Moreno Sastre, insisted, "It's not just the collective imagination." He told *Wall Street Journal* reporters that witnesses numbered in the "thousands." A priest, Father Manuel Prados Munoz of the mountain village of Maria, claimed repeated sightings, sometimes as many as a dozen a month. He said the planes would show up whenever his desktop barometer—and his eyes—registered that a storm was coming. (A barometer is an instrument for measuring atmospheric pressure, usually indicated by the movement of a column of mercury in a sealed glass tube. A storm is expected when the barometer is falling quickly; when it is rising, fair weather is predicted.) When local people began to report their sightings of planes to Munoz, he learned of hundreds of similar cases.

With these sightings, no explanation really makes sense. There was absolutely no evidence that a supersecret military or other weather-control operation was the cause. And happily for all concerned, when rain ended the droughts, the mystery planes—real or imagined—passed out of sight and soon out of mind.

Sources:

Corliss, William R., ed., *Handbook of Unusual Natural Phenomena,* Glen Arm, Maryland: The Sourcebook Project, 1977.

Corliss, William R., ed., *Tornados, Dark Days, Anomalous Precipitation, and Related Weather Phenomena: A Catalog of Geophysical Anomalies,* Glen Arm, Maryland: The Sourcebook Project, 1983.

Michell, John, and Robert J. M. Rickard, *Phenomena: A Book of Wonders,* New York: Pantheon Books, 1977.

BALL LIGHTNING

At 6:30 P.M. on October 8, 1919, at a busy downtown street crossing in Salina, Kansas, a "ball of fire as large as a washtub floating low in the air" struck the side of a building, ripped out bricks, and destroyed a second-story window.

Just moments after hearing what sounded like thunder, a Parisian man reported an extraordinary sight: a fireball the size of a human head emerged from the fireplace of his fourth-story apartment and darted toward him "like a cat." The man quickly raised his feet, and the ball moved to the center of the room. Though bright, it gave off no noticeable heat. It rose slightly, headed back to the fireplace, and retreated up the chimney, exploding just before it escaped into the open air. The chimney top received a fair amount of damage.

This 1852 account describes ball lightning, a strange—and so far unexplained—natural occurrence. While some scientists believe that ball lightning is real, others question it, because reliable data is rare, with most evidence consisting of hearsay reports and questionable photographs. More bothersome still is the fact that no known scientific principle can account for it!

Some doubters of ball lightning feel that witnesses are reporting optical illusions, visual afterimages caused by watching something very bright, like a flash of lightning. Others believe that observers are experiencing a natural phenomenon like St. Elmo's fire, the "halo" of electricity sometimes discharged from an object projecting above ground during an electrical storm.

While the second explanation seems a reasonable one, James Dale Barry, a leading scientific expert on the subject, pointed out that St. Elmo's fire cannot move about like ball lightning, explaining, "It may move along a conductor, sometimes pulsating as it moves, but it does not free itself from the conductor. Thus, it does not exhibit the descending, hovering, or flying motions that are common to ball lightning."

The first investigator to describe ball lightning in scientific literature was the Russian G. W. Richman. Tragically, his study led to his death. In 1754, during a thunderstorm, Richman was trying to measure the energy of a lightning strike. As he stood behind his equipment, a small, blue, fist-sized ball came out of the electrodes and floated toward his face. A moment later it exploded, killing him and knocking his assistant unconscious.

Fortunately, deaths related to ball lightning are rare, but many observers have witnessed its destructive power. During an electrical storm in Paris in July 1849, a red ball hung about 20 feet above a tree. The tree suddenly caught fire, burned, and burst open, jagged streaks of lightning shooting out in all directions. One hit a nearby house and blew a cannon-size hole in it. What remained of the lightning ball started to spin and spark and then exploded with great force, knocking down three people nearby.

At 6:30 P.M. on October 8, 1919, at a busy downtown street crossing in Salina, Kansas, a "ball of fire as large as a washtub floating low in the air" struck the side of a building, ripped out bricks, and destroyed a second-story window. It then exploded with a "bang that resembled the noise made by the discharge of a large pistol, filling the air with balls of fire as large as baseballs, which floated away in all directions," reported that month's issue of *Monthly Weather Review*. "Some of these balls followed trolley and electric-light wires in a snaky sort of manner and some simply floated off through the air independently of any objects

In 1852 a Parisian man reported that a fireball emerged from his fireplace and darted toward him "like a cat."

Fireball seen during a storm in Salagnac, France, 1845, possibly ball lightning. From Camille Flammarion, *L'Atmosphère,* 1888.

near by. An electric switch box across the street was ripped open and a transformer destroyed, leaving the east side of the town in darkness."

In the summer of 1960, as Louise Matthews of south Philadelphia lay on her living room couch, she looked up to see a huge red ball coming through a window and its blinds, both closed and neither disturbed by the entry. When the globe, which made a sizzling sound, passed by her, Mrs. Matthews felt a tingling on the back of her neck. She put her hand to the spot but felt nothing. The ball went through the living room and into the dining room, leaving—again without damage—through another closed window. She called her husband, who came home from work to find the back of her hand burned. And the hair at the back of her head had fallen out, leaving the skin there as smooth as that on her face.

During a violent early-evening thunderstorm on August 12, 1970, a "red ball of fire" appeared above Sidmouth, England, crackled for a few seconds, then exploded with a deafening roar. Jagged flashes of lightning shot from it toward the ground. At that moment, 2,500 area television sets were cut off.

Despite the nature of these accounts, ball lightning is not always harmful, nor does it always explode at the end of its run. In a November 1930 issue of *Nature,* British scientist Alexander Russell recalled a sighting where two balls of lightning acted very differently. "One of them struck a building and burst with a loud report, causing inhabitants to open the windows and look out to see what had happened," he related, while "the other drifted slowly away." The scientist added that when ball lightning drifts about it makes a slight noise—like "the purring of a cat."

More Theories

Much of the problem behind explaining ball lightning is the result of the varying descriptions witnesses have given. The ball either explodes loudly or disappears silently; it is white, orange, red, blue, or purple; it is small or large; it lasts for a few seconds or several minutes. Science writer Gordon Stein remarked that while these differences may seem minor, "they cause theorists no end of difficulties. Explanations that will work for a ball of one second's duration, for example, cannot account for a 10-second ball," for a ball that lasts one minute or more "requires an energy content so high that there is no known way for it to be formed."

Also making an explanation tricky is ball lightning's ability to penetrate the metal walls of in-flight aircraft. On March 19, 1963, R. C. Jennison, a professor of electrical energy, saw a globe of lightning first outside, then inside, an airliner he was taking from New York to Washington. An electrical storm was in progress. According to Stein, the incident casts doubt on several theories about ball lightning because "microwave, electric, radio or heat energy ... could not have gotten through the metal fuselage [body of the plane]." He also felt that it disproved the ideas of E. T. F. Ashby and C. Whitehead, who suggested that ball lightning consisted of antimatter. "When antimatter—matter composed of the counterparts of ordinary matter—comes in contact with normal matter, both are annihilated," Stein wrote. "Antimatter would have a difficult time getting through the body or window of an airplane without colliding with some regular matter, thus destroying itself."

Sources:

Barry, James Dale, *Ball Lightning and Bead Lightning: Extreme Forms of Atmospheric Electricity,* New York: Plenum Press, 1980.

Constance, Arthur, *The Inexplicable Sky,* New York: The Citadel Press, 1957.

Evans, Hilary, ed., *The Frontiers of Reality: Where Science Meets the Paranormal,* Wellingborough, Northamptonshire, England: The Aquarian Press, 1989.

SKYQUAKES

As unlikely as it
seemed in this wild
country, the noises
resembled heavy
cannon fire!

While exploring the Rocky Mountains on July 4, 1808, members of the Lewis and Clark expedition heard strange noises coming from some distant mountains. They were heard at different times of the day and night, ringing out one at a time or in five or six loud bursts. Weather didn't seem to affect the sounds, for they were often heard when the sky was cloudless and the wind was still. As unlikely as it seemed in this wild country, the noises resembled heavy cannon fire!

An 1829 exploration party along the Darling River, near what is now Bourke, New South Wales, Australia, was also puzzled one afternoon by the sound of heavy gunfire in the unsettled territory. The expedition leader recorded this in his diary on February 7:

> the day had been remarkably fine, not a cloud was there in the heavens, nor a breath of air to be felt. On a sudden we heard what seemed to be the report of a gun fired at the distance of between five and six miles. It was not the hollow sound of an earthly explosion, or the sharp cracking noise of falling timber, but in every way resembled a discharge of a heavy piece of ordnance [cannon]. On this all were agreed, but no one was certain where the sound proceeded.... I sent one of the men immediately up a tree, but he could observe nothing unusual.

It is likely that the faraway gunfire that both expeditions heard (but could not locate) was the oddity known as skyquakes. Described as muffled thunder or cannon fire, skyquakes have been heard all over the world. One instance of skyquakes takes place in India and is known as the Barisal Guns; another example in Connecticut is called the Moodus Noises.

Theories

In past centuries witnesses to the strange sounds have attempted to record and explain them in their folklore and legends; in our time, imaginative writers have even linked them to UFOs or parallel universes. Scientists, however, offer a more sensible explanation: skyquakes are not really sounds from the sky but come, instead, from under the ground. Earthquakes and other earthly vibrations—sometimes so small that only an instrument can measure them—create these odd noises; it is because the sounds come from so deep underground that confusion follows—they *do* seem to be made in the sky.

While scientists don't know why skyquakes are heard in some places and not others, they are working hard to find out. Since 1981 the Weston Observatory of Boston College has been studying earthquake activity in New England and its connection to the Moodus Noises. Perhaps sometime soon they will discover exactly what causes skyquakes.

Did the Lewis and Clark expedition experience skyquakes in the Rocky Mountains?

Sources:

Corliss, William R., *Handbook of Unusual Natural Phenomena,* Glen Arm, Maryland: The Sourcebook Project, 1977.
Fort, Charles, *The Books of Charles Fort,* New York: Henry Holt and Company, 1941.

TUNGUSKA EVENT

In 1908 a strange explosion took place in a remote swampy area of Siberia, Russia. Today, nearly a century later, the event is still the source of wonder, argument, and a number of explanatory theories.

At 7:15 on the morning of June 30, a blazing white light was seen falling over the forests northwest of Lake Baykal near the Stony Tunguska River. It was so bright that it cast shadows on the earth beneath it. As it fell, it flattened trees and smashed houses, finally exploding with such force that seismic shocks (earth vibrations) were recorded around the world. A giant "pillar of fire" rose straight up and was seen hundreds of miles away. As huge thunderclaps sounded, a blistering current of hot air tore through the area, setting off fires in forests and towns. At least three shock waves followed the thermal (heat) wave. The destruction was massive, extending for 375 miles. Thick, dark clouds rose above the site of the explosion, and a black rain of dirt and particles fell on central Russia. That night the sky remained eerily bright all through northern Europe.

Perhaps an explanation for this gigantic explosion would have been easy to find had scientists been able to go to the site immediately. But because Russia was suffering from grave political upheaval—which would soon erupt into war and revolution—scientists focused on other concerns. The first expedition to the area, led by Leonid Kulik of the Russian Meteorological Institute, did not take place until 13 years later.

Expedition members expected to come upon a huge meteorite crater but were surprised to find nothing of the sort. Instead, they discovered that trees there had been damaged from above. Moreover, those closest to the impact site were still standing, though missing bark and branches. Trees farther away were flattened and pointing out from the site. Kulik and his companions searched carefully but could find no meteorite fragments.

Nonetheless, Kulik, whose work continued until the outbreak of World War II (in which he was killed), remained certain that a meteorite was the cause of the explosion. A colleague, Vasili Sytin, disagreed, because there was no evidence that indicated a source outside the earth's atmosphere. Instead, he suggested an earthly explanation: an unusually violent windstorm.

> That night the sky remained eerily bright all through northern Europe.

METEOROIDS

A *meteoroid* is any piece of matter—ranging in mass from a speck of dust to thousands of tons—that travels through space; it is composed largely of stone or iron or a mixture of the two. When a meteoroid enters the earth's atmosphere it becomes visible and is called a *meteor*. A *meteorite* is a meteor that survives the fall to earth.

Extraterrestrial Explanations

After learning of the destruction left by atomic bomb attacks on the Japanese cities of Hiroshima and Nagasaki at the end of World War II, Soviet science-fiction writer A. Kasantsev published a story about Tunguska in the January 1946 issue of *Vokrug Sveta*. In his story a Martian spaceship—vaporized in a nuclear explosion—causes the destruction of Tunguska.

Though the story was only a fantasy, Kasantsev was attacked by Soviet scientists for presenting what they believed was a ridiculous explanation. Nonetheless, the idea that a spaceship had exploded over Tunguska caught on, capturing the public's imagination—first in the Soviet Union, then around the world. Two Soviet scientists, Felix Zigel and Aleksey Zolotov, were strong supporters of the theory in later decades, with Zolotov even claiming to have detected "abnormal radioactivity" at the Tunguska site. Testing by other scientists, however, provided no solid proof. A popular 1976 book, *The Fire Came By*, defended the spaceship idea, but most Western—and Russian—scientists continued to reject the theory.

1929 photograph of trees devastated by the Tunguska catastrophe of 1908.

A 1910 photo of
Halley's Comet.

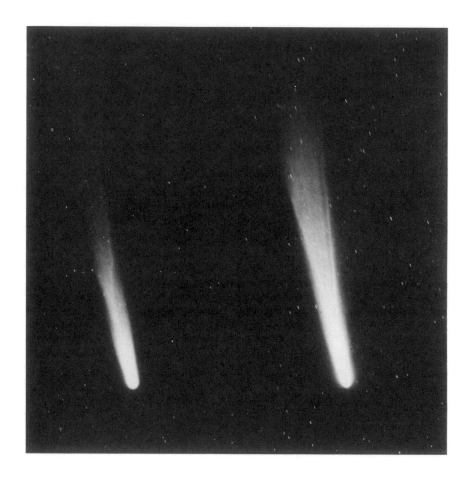

COMETS

A *comet* is a small heavenly body consisting mostly of gases, the movement of which depends on the sun's gravity. As a comet approaches the sun, particles and gases are driven off, often to form a tail, which can extend up to 100 million miles.

Today explanations for the Tunguska event have refocused on meteorites, comets, and asteroids. Almost all scientists agree that the object, possibly as large as 200 yards around, never struck the earth but exploded in midair because of the air pressure that built up beneath it as it fell.

"Had the Tunguska object been a comet," wrote Stephen P. Maran in *Natural History,* "the failure to find fragments of rock or iron from the explosion would be understandable. Any cometary ice that reached the ground probably would have melted before the first scientific expedition reached the site.... [If] the Tunguska object was an asteroid or meteoroid and thus made of stone and iron ... either there are fragments, which have been overlooked by repeated Soviet scientific expedites, or ... the incoming object ... shattered totally into dust in the explosion."

Sources:

Baxter, John, and Thomas Atkins, *The Fire Came By: The Riddle of the Great Siberian Explosion,* Garden City, New York: Doubleday and Company, 1976.

Ganapathy, Ramachandran, "The Tunguska Explosion of 1908: Discovery of Meteoritic Debris Near the Explosion Site and at the South Pole," *Science* 220,4602, June 10, 1983, pp. 1158-61.

Maran, Stephen P., "What Struck Tunguska?," *Natural History* 93,2, February 1984, pp. 36-37.

FURTHER INVESTIGATIONS

BOOKS

Alien Contacts

Adamski, George, *Inside the Space Ships,* **New York: Abelard-Schuman, 1955.**

In 1952 Adamski, a UFO writer and photographer of space ships, reported that he met a flying saucer pilot from Venus. He then embarked on a colorful career as a contactee with connections on Mars, Venus, and Saturn. In 1954 a Venusian scoutcraft allegedly flew Adamski around the moon, and this book details his lunar odyssey.

Hopkins, Budd, *Intruders: The Incredible Visitations at Copley Woods,* **New York: Random House, 1987.**

Hopkins is best known for his UFO-abduction reports, which he, more than any other writer or investigator, has brought to wide public attention. *Intruders,* like his 1981 *Missing Time,* recounts the stories (many evoked under hynosis) of witnesses who were abducted by large-headed, gray-skinned humanoids.

Strieber, Whitley, *Communion: A True Story,* **New York: Beach Tree/William Morrow, 1987.**

The best-selling UFO book of all time recounts the author's experiences with "visitors"—small, almond-eyed, gray-skinned humanoid occupants of UFOs. A fairly well-known writer of Gothic and futuristic fiction, Strieber contacted UFO-abduction investigator Budd Hopkins after a strange but barely remembered alien contact experience. He wrote this book after hypnosis, which revealed several visitor-related events in his life. William Morrow paid

Strieber $1 million for the book, which attracted enormous attention and a huge reading audience. The film version, starring Christopher Walken as Strieber, met with a fairly tepid response.

Ancient Astronauts

Temple, Robert K. G., *The Sirius Mystery*, New York: St. Martin's Press, 1977.

If the ancient astronaut fad of the 1970s produced one book of substance, many agree this is the one. Learned and extensively researched, it presents a complex, many-sided argument for an early extraterrestrial presence in West Africa.

Von Däniken, Erich Anton, *Chariots of the Gods?: Unsolved Mysteries of the Past*, New York: G. P. Putnam's Sons, 1970.

Along with several other writers in the 1960s, Swiss writer von Däniken theorized in this best-selling book that the gods of Judaism, Christianity, and other religions were extraterrestrials who, through direct interbreeding with our primitive ancestors or through direct manipulation of the genetic code, created *Homo sapiens*. These beings were also responsible for the archaeological and engineering wonders of the ancient world, as well as the mysterious Nazca lines. By no means the original text on the ancient astronaut theory, von Däniken's book took the world by storm, spawning a multitude of sequels, other books, and a popular film on the topic.

Anomalies, general

Fort, Charles, *The Books of Charles Fort*, New York: Henry Holt and Company, 1941.

Until Fort, the pioneer of unexplained physical phenomena, began his extensive research into anomalies, no one knew how "ordinary" and frequent strange happenings really were. This 1941 collection contains *Book of the Damned* (1919), a highly celebrated book that first exposed the reading public to giant hailstones, red and black rains, falls from the sky, unidentified flying objects, and other anomalies. Also in the volume are *New Lands* (1923), *Lo!* (1931), and *Wild Talents* (1932). Along with collecting and recording anomalies, these books present Fort's famously outlandish "theories," which he satirically regarded as no less preposterous than those scientists were offering to explain anomalies.

For further investigation of general anomalies see:

Bord, Janet, and Colin Bord, *Unexplained Mysteries of the 20th Century,* Chicago: Contemporary Books, 1989.
Cohen, Daniel, *The Encyclopedia of the Strange,* New York: Dorset Press, 1985.
Coleman, Loren, *Curious Encounters: Phantom Trains, Spooky Spots and Other Mysterious Wonders,* Boston: Faber and Faber, 1985.

Corliss, William R., ed., *Handbook of Unusual Natural Phenomena,* Glen Arm, MD: The Sourcebook Project, 1977.

Knight, Damon, *Charles Fort: Prophet of the Unexplained,* Garden City, NY: Doubleday and Company, 1970.

Michell, John, and Robert J. M. Rickard, *Living Wonders: Mysteries and Curiosities of the Animal World,* London: Thames and Hudson, 1982.

Bermuda Triangle

Berlitz, Charles, with J. Manson Valentine, *The Bermuda Triangle,* Garden City, NY: Doubleday and Company, 1974.

The Bermuda Triangle fever peaked with the publication of this best-selling book, which sold over five million copies worldwide. Like most of the Triangle books, there is little evidence of original research in its account of the disappearances of planes and boats off the Florida coast, and many of the "facts" that created the mystery were later discredited.

Kusche, Lawrence David, *The Bermuda Triangle Mystery—Solved,* New York: Harper & Row, 1975.

A thorough debunking of what Kusche calls the "manufactured mystery" of the Bermuda Triangle disappearances. For this book, Kusche did the research other Triangle writers had neglected. Weather records, newspaper accounts, official investigators' reports, and other documents indicated that previous Triangle writers had played fast and loose with the evidence.

Cattle Mutilations

Kagan, Daniel, and Ian Summers, *Mute Evidence,* New York: Bantam Books, 1984.

The authors traveled extensively through the western United States and Canada researching the bizarre stories of cattle mutilations. This definitive account exposes journalistic sensationalism and mass hysteria as the only solid basis for the modern-day myth.

Crop Circles

Delgado, Pat, and Colin Andrews, *Circular Evidence,* London: Bloomsbury, 1989.

Delgado, Pat, and Colin Andrews, *Crop Circles: The Latest Evidence,* London: Bloomsbury, 1990.

Two best-selling books on the English phenomenon. The authors, who are active crop circle investigators and founders of the Circles Phenomenon Research group, do not believe that the scientific explanations offered for the mystery so far—either weather- or hoax-related—can begin to explain the anomaly.

Cryptozoology

Bord, Janet, and Colin Bord, *Alien Animals,* **Harrisburg, PA: Stackpole Books, 1981.**

> The best of the Bords' books on paranormal (supernatural) cryptozoology, *Alien Animals* expresses the authors' theory that all mysterious animal sightings, along with UFO sightings and other unexplainable apparitions, are manifestations of "a single phenomenon."

Heuvelmans, Bernard, *On the Track of Unknown Animals,* **New York: Hill and Wang, 1958.**

> Heuvelmans, who is considered the father of cryptozoology, collected innumerable printed references to mysterious, unknown animals from scientific, travel, and popular literature and put them together in this large, informative book that sold over a million copies around the world.

Michell, John, and Robert J. M. Rickard, *Living Wonders: Mysteries and Curiosities of the Animal World,* **New York: Thames and Hudson, 1982.**

> In this lively and literate book, Rickard, founder of the *Fortean Times,* and Michell capture Charles Fort's rich humor and sense of cosmic comedy, while providing encyclopedic, world-ranging coverage of current and historic anomalies.

For further investigation of cryptozoology see:

Caras, Roger A., *Dangerous to Man: The Definitive Story of Wildlife's Reputed Dangers,* New York: Holt, Rinehart and Winston, 1975.

Clark, Jerome, and Loren Coleman, *Creatures of the Outer Edge,* New York: Warner Books, 1978.

Mackal, Roy P., *Searching for Hidden Animals: An Inquiry into Zoological Mysteries,* Garden City, NY: Doubleday and Company, 1980.

Skuker, Karl P. N., *Mystery Cats of the World: From Blue Tigers to Exmoor Beasts,* London; Robert Hale, 1989.

South, Malcolm, ed., *Mythical and Fabulous Creatures: A Source Book and Research Guide,* New York: Greenwood Press, 1987.

Extinct Animal Sightings

Mackal, Roy P., *A Living Dinosaur?: In Search of Mokele Mbembe,* **New York: E. J. Brill, 1987.**

> After two expeditions to the Congo to investigate reports of the mysterious mokele mbembe, University of Chicago biologist Roy Mackal persuasively argues in this book that the monster so frequently reported in this remote area of Africa is in fact some form of sauropod, a dinosaur thought to have been extinct for millions of years.

For further investigation of extinct animal sightings see:

Doyle, Sir Arthur Conan, *The Lost World* (fiction), London: Hodder and Stoughton, 1912.

Guiler, Eric R., *Thylacine: The Tragedy of the Tasmanian Tiger,* Oxford, England: Oxford University Press, 1985.

Folklore

Benwell, Gwen, and Arthur Waugh, *Sea Enchantress: The Tale of the Mermaid and Her Kin,* New York: The Citadel Press, 1965.

This highly regarded book on merfolk lore examines the many traditions of merfolk sightings throughout history and concludes that merfolk must be some kind of unknown, unrecorded species of sea animal.

Evans-Wentz, W. Y., *The Fairy-Faith in Celtic Countries,* New York: University Books, 1966.

The author, an anthropologist of religion, traveled throughout the British Isles recording oral traditions of fairy belief. The resulting book, a classic in folklore studies, also presents the author's extensive research on the existence of "such invisible intelligences as gods, genii, daemons, all kinds of true fairies, and disembodied men."

For further investigation of folklore see:

Otten, Charlotte F., ed. *A Lycanthropy Reader: Werewolves in Western Culture,* New York: Dorset Press, 1986.

Government Cover-ups

Moore, William L., with Charles Berlitz, *The Philadelphia Experiment: Project Invisibility—An Account of a Search for a Secret Navy Wartime Project That May Have Succeeded—Too Well,* New York: Grosset and Dunlap, 1979.

A popular book about the bizarre Philadelphia Experiment during World War II when, according to the letters of a very questionable character, a ship was made invisible and instantaneously transported between two docks, causing its crew members to become insane. The unflagging interest of three Office of Naval Research officers with the far-fetched story was the most unexplainable mystery here. Moore's book inspired the 1984 science fiction movie.

Randle, Kevin D., and Donald R. Schmitt, *UFO Crash at Roswell,* New York: Avon Books, 1991.

The government cover-up of the 1947 "Roswell incident" was effective enough to submerge the story of the crashed flying saucer for nearly 30 years, but the investigation is not closed. Randle and Schmitt have joined others in collecting the testimony of hundreds of witnesses, from local ranchers to air force generals. From these reports they have reconstructed a complex series of events, and the Roswell incident has become one of the best-documented cases in UFO history.

Hairy Bipeds

Napier, John, *Bigfoot: The Yeti and Sasquatch in Myth and Reality,* New York: E. P. Dutton and Company, 1973.

A scientist's view of the abundant evidence of Bigfoot's existence. Napier, a primatologist and the curator of the primate collections at the Smithsonian Institution, was one of the very few conventional scientists to pay serious attention to Bigfoot and other hairy bipeds.

Sanderson, Ivan T., *Abominable Snowmen: Legend Come to Life*, Philadelphia, PA: Chilton Book Company, 1961.

The first book to discuss Bigfoot/Sasquatch in any comprehensive manner. Sanderson linked the North American sightings with worldwide reports of "wild men," Almas, and yeti. An encyclopedic view of hairy biped traditions.

For further investigation of hairy bipeds see:

Bord, Janet, and Colin Bord, *The Bigfoot Casebook,* Harrisburg, PA: Stackpole Books, 1982.
Byrne, Peter, *The Search for Big Foot: Monster, Myth or Man?*, Washington, DC: Acropolis Books, 1976.

Lake Monsters

Zarzynski, Joseph W., *Champ: Beyond the Legend*, Port Henry, NY: Bannister Publications, 1984.

Vermont's own version of Nessie has a strong advocate in Zarzynski, who formed the Lake Champlain Phenomena Investigation in the 1970s for extensive research of historical sightings and surveillance of the lake. The author links Champ with the Loch Ness monster in this authoritative, if speculative, book.

Loch Ness Monsters

Dinsdale, Tim, *Loch Ness Monster,* 4th edition, Boston: Routledge and Kegan Paul, 1982.

Over a period of 27 years, British aeronautical engineer Tim Dinsdale made 56 separate expeditions to Ness and spent 580 days watching for the animals. In all, he had three sightings, one of which he filmed. The Dinsdale film is still considered compelling evidence of the existence of the monsters. His highly regarded book, *Loch Ness Monster,* went through four editions between 1961 and 1982.

Holiday, F. W., *The Dragon and the Disc: An Investigation into the Totally Fantastic,* New York: W. W. Norton and Company, 1973.

The author, the most radical of the Nessie theorists, originally suggested that the animals in Loch Ness were enormous prehistoric slugs. In this book he changed to an explicitly occult interpretation: Nessies are dragons in the most literal, traditional sense—they are supernatural and evil.

Mackal, Roy P., *The Monsters of Loch Ness,* Chicago: The Swallow Press, 1976.

Mackal was the scientific director of the Loch Ness Phenomena Investigation Bureau from 1965 to 1975 and this book, which grew out of his field work, is a cryptozoological classic.

For further investigation of Loch Ness monsters see:

Bauer, Henry H., *The Enigma of Loch Ness: Making Sense of a Mystery*, Urbana, IL: University of Illinois Press, 1986.
Binns, Ronald, *The Loch Ness Mystery Solved,* Buffalo, NY: Prometheus Books, 1984.

Sea Monsters

Heuvelmans, Bernard, *In the Wake of the Sea-Serpents,* New York: Hill and Wang, 1968.

In the most comprehensive volume ever written on the sea serpent, Heuvelmans analyzes 587 sea-serpent reports. He considers 358 of these to be authentic sightings, 49 hoaxes, 52 mistakes, and the rest lack sufficient detail to analyze. The author theorizes that the term "sea serpent" actually covers several unrecognized marine animals, which he classifies in the conclusive chapter.

Sanderson, Ivan T., *Invisible Residents: A Disquisition upon Certain Matters Maritime, and the Possibility of Intelligent Life under the Waters of This Earth,* New York: World Publishing Company, 1970.

Zoologist Sanderson demonstrates his wide-ranging curiosity and his creative imagination in this book in which he theorizes that an intelligent, technologically advanced civilization lives, undetected by the rest of us, in the oceans of the earth. These beings may be extraterrestrials, according to Sanderson, and some UFOs may be their versatile submarines.

For further investigation of sea monsters see:

Lester, Paul, *The Great Sea Serpent Controversy: A Cultural Study*, Birmingham, England: Protean Publications, 1984.

Unidentified Airships

Cohen, Daniel, *The Great Airship Mystery: A UFO of the 1890s,* New York: Dodd, Mead, and Company, 1981.

An entertaining, informative look at the famous turn-of-the-century UFO wave.

UFOs

Hynek, J. Allen, *The UFO Experience: A Scientific Inquiry,* Chicago: Henry Regnery Company, 1972.

At one time a consultant to the air force's UFO-debunking mission, astronomer Hynek changed from a skeptical attitude to a solid belief in the reality of UFOs. In his well-received book, *The UFO Experience,* he blasts the air force's UFO evidence-debunking projects and argues persuasively that science would be greatly furthered by an open-minded study of the subject.

Ruppelt, Edward J., *The Report on Unidentified Flying Objects,* Garden City, NY: Doubleday and Company, 1956.

When Lieutenant Ruppelt, an intelligence officer in the air force, took over the air force investigations of UFOs in the early 1950s, he insisted that investigations be carried out without prior judgments about the reality or unreality of UFOs. By the time he left the project two years later, Ruppelt was largely convinced that space visitors did exist. This memoir of his experiences is considered one of ufology's most important books.

Vallee, Jacques, *Passport to Magonia: From Folklore to Flying Saucers,* Chicago: Henry Regnery Company, 1969.

Vallee is one of the leading theorists on UFOs. *Passport to Magonia* marks his shift from scientific examination of UFO evidence to theories that UFO phenomena have their origins in another reality that is beyond the bounds of scientific analysis. Vallee proposes that inquirers need to immerse themselves in traditional supernatural beliefs—in fairies, gods, and other fabulous beings—in order to begin to understand that aliens and UFOs are only the modern manifestation of ancient beings.

For further investigation of UFOs see:

Blevins, David, *Almanac of UFO Organizations and Publications,* 2nd edition, San Bruno, California: Phaedra Enterprises, 1992.

Clark, Jerome, *The Emergence of a Phenomenon: UFOs from the Beginning through 1959— The UFO Encyclopedia,* Volume 2, Detroit, MI: Omnigraphics, 1992.

Clark, Jerome, *UFOs in the 1980s: The UFO Encyclopedia,* Volume 1, Detroit, MI: Apogee Books, 1990.

Weather Phenomena

Corliss, William R., ed., *Handbook of Unusual Natural Phenomena,* Glen Arm, MD: The Sourcebook Project, 1977.

Corliss, William R., ed., *Strange Phenomena,* two volumes, Glen Arm, MD: The Sourcebook Project, 1974.

Corliss, William R., ed., *Tornados, Dark Days, Anomalous Precipitation, and Related Weather Phenomena: A Catalog of Geophysical Anomalies,* Glen Arm, MD: The Sourcebook Project, 1983.

Corliss, a physicist who systematically catalogued more than 20 volumes' worth of anomalies, applies a conservative, scientific approach to bizarre and seemingly unaccountable events. He was particularly interested in unusual weather, and these volumes of his monumental Sourcebook Project are an invaluable resource in this area.

PERIODICALS

FATE

Llewellyn Worldwide, Ltd.
Box 64383
St. Paul, Minnesota 55164

Monthly

Fate was created in 1948 by Ray Palmer, science fiction editor of *Amazing Stories* and *Fantastic Adventures*, and Curtis Fuller, editor of *Flying*. Floods of flying saucer reports swept the nation following Kenneth Arnold's June 24, 1947, sighting of nine fast-moving discs. Palmer had found in his work that these kinds of "true mystery" stories were wildly popular, and, since no mass-circulation periodical devoted to such matters existed, the two editors decided to fill the void in the market.

Fate covered mysteries relating to ufology, cryptozoology, and archaeology, but its greatest emphasis was on psychic phenomena. In the 1950s Palmer sold his share of the magazine to Fuller, who then edited it with his wife, Mary Margaret Fuller. The magazine was the most successful popular psychic magazine of all time, and achieved a peak circulation of 175,000 during the late 1970s. In 1988 Mary Margaret Fuller was replaced as editor by Jerome Clark, and Phyllis Galde later took over. *Fate*'s 500th issue was published in November 1991.

Strange Magazine

Box 2246
Rockville, Maryland 20852

Semiannual

In the introduction to the first issue of *Strange Magazine,* Mark Chorvinsky, the magician and filmmaker who created the publication, wrote: "We range from wild theoretical speculation to cautious skeptics—including every shade of worldview in between. Some of us are philosophers, others investigators and researchers—surrealistic scientists who catalog the anomalous, the excluded, the exceptional." The focus of *Strange* is on physical rather than psychic anomalies, and it includes topics such as cryptozoology, ufology, archaeological mysteries, falls from the sky, crop circles, and behavioral oddities. Chorvinsky, who has a particular interest in hoaxes, has exposed a number of dubious claims, most notably those associated with English magician and trickster Tony "Doc" Shiels, whose widely reproduced photographs of the Loch Ness monster and a Cornish sea serpent had many fooled. *Strange* reflects its editor's attitude toward anomalous phenomena: open-minded but not credulous.

Published semiannually, each issue of *Strange* is 64 pages long, full of lively graphics and well-written, well-researched articles. It is essential reading for all committed anomalists.

MAJOR ORGANIZATIONS
AND THEIR PUBLICATIONS

Ancient Astronaut Society

1821 St. Johns Avenue
Highland Park, IL 60035

Bimonthly bulletin: *Ancient Skies*

The Ancient Astronaut Society was formed in 1973 and is based on the belief that advanced space beings visited the earth early in human's history and possibly played a part in the development of human intelligence and technology. The organization is directed by attorney Gene M. Phillips in Chicago. European director Erich von Däniken, who wrote many books about the idea—including the wildly popular *Chariots of the Gods?*—operates out of Switzerland. The organization publishes a bimonthly bulletin, *Ancient Skies,* in both English and German. It also meets in a different world city every year and sponsors archaeological expeditions to sites where ancient marvels, viewed as evidence for the group's beliefs, can be seen by members firsthand.

Center for Scientfic Anomalies Research (CSAR)

Box 1052
Ann Arbor, MI 48106

Journal: *Zetetic Scholar*

Formed in 1981, the Center for Scientific Anomalies Research is a "private center which brings together scholars and researchers concerned with furthering responsible scientific inquiry into and evaluation of claims of anomalies and the paranormal." Director Marcello Truzzi and associate director Ron Westrum are sociologists of science at Eastern Michigan University in Ypsilanti. From 1978 to 1987 Truzzi, who had cofounded and then resigned from the Committee for the Scientific Investigation of Claims of the Paranormal (see CSICOP entry below), edited the journal *Zetetic Scholar,* a forum in which believers and nonbelievers could discuss and debate their views on anomalous subjects. CSAR was created as a parent organization for the publication; a number of important physical and biological scientists, psychologists, and philosophers are among its consultants.

Committee for the Scientific Investigation
of Claims of the Paranormal (CSICOP)

Box 703
Buffalo, NY 14226

Quarterly magazine: *Skeptical Inquirer*

The Committee for the Scientific Investigation of Claims of the Paranormal was formed in 1976 by Paul Kurtz, a professor of philosophy at the State University of New York at Buffalo, and science sociologist Marcello Truzzi (see CSAR entry above). Unfortunately, the two founders had different aims for the organization right from the start. Kurtz and his followers were rigid disbelievers in ufology, astrology, and other subjects that existed on the fringes of science; more than just skeptics, they viewed such unusual ideas as threats to reason and civilization. Truzzi felt that at least some claims, especially those made by serious parapsychologists, cryptozoologists, and ufologists, were worth investigating. Truzzi had hoped that the organization would practice fairminded *nonbelief* rather than ridiculing *disbelief.* He resigned a year later.

From the beginning CSICOP attracted many famous scientists. The organization was well funded and by the late 1980s its quarterly magazine, the *Skeptical Inquirer,* would claim a circulation of more than 30,000—the world's second most popular magazine on anomalies and the paranormal (after the psychic digest *Fate*). CSICOP sponsors regular conferences. Through its connected publishing house, Prometheus Books, it releases works expressing the debunker's view of UFOs, the Loch Ness monster, the Bermuda Triangle, and other "antiscientific" matters.

Fund for UFO Research, Inc. (FUFOR)

Box 277
Mount Rainier, MD 20712

The Fund for UFO Research was founded in 1979 and provides grants for scientific research and educational projects on UFO-related subjects. Studies have included investigations of UFO photographs and crash reports. The organization publishes findings from these projects from time to time. Funds are granted by a ten-member board of directors of scientists and scholars.

International Fortean Organization (INFO)

Box 367
Arlington, VA 22210

Quarterly journal: *INFO Journal*

The International Fortean Organization was founded in 1965 by Ronald J. Willis. It is dedicated to the memory and interests of pioneering anomaly collector Charles Fort (1874-1932). The *INFO Journal* is a forum for a wide variety of unexplained physical happenings, both past and present. The organization sponsors a yearly "FortFest" in the Washington, D.C., area, where well-known writers and researchers discuss anomalies.

INFO is a successor to the Fortean Society (1931-1960). Under the direction of writer and advertising man Tiffany Thayer, that organization and its *Fortean Society Magazine* (retitled *Doubt* in 1944) were known for their weird ideas and eccentric writers. The society oversaw the important publication of *The Books of Charles Fort,* a collection of the anomalist's writings, in 1941.

International Society of Cryptozoology (ISC)

Box 43070
Tucson, AZ 85733
Yearly journal: *Cryptozoology*
Quarterly newsletter: *ISC Newsletter*

The International Society of Cryptozoology was formed in early 1982. At its founding meeting at the Museum of Natural History of the Smithsonian Institution in Washington, D.C., the organization defined its purpose as the "scientific inquiry, education, and communication among people interested in animals of unexpected form or size, or unexpected occurrence in time and space." Roy P. Mackal, a University of Chicago biologist, and University of Arizona ecologist J. Richard Greenwell had worked behind the scenes for more than a year to put the organization together. (They, along with cryptozoology pioneer Bernard Heuvelmans, became the ISC's elected officers.)

With its serious, scientific approach to the subject of "unexpected" animals, the organization has been able to attract well-respected zoologists, anthropologists, and others as members. The ISC holds an annual meeting, always at a university or scientific institute. It publishes the quarterly *ISC Newsletter* and the yearly journal *Cryptozoology*. Although mainstream science has still not fully accepted cryptozoology, the ISC has enhanced the respectability of the field.

Intruders Foundation (IF)

Box 30233
New York, NY 10011
Yearly bulletin: *IF: The Bulletin of the Intruders Foundation*

The Intruders Foundation was created by Budd Hopkins, author of two popular books on UFO abductions. The purpose of the organization is to fund research and to offer therapeutic help to the many people who have contacted Hopkins, disturbed by their own claimed abduction experiences. The foundation has an informal national network of mental health professionals who volunteer to counsel these people. *IF: The Bulletin of the Intruders Foundation* appears yearly, and discusses abduction cases, investigations, and other related matters.

J. Allen Hynek Center for UFO Studies (CUFOS)

2457 West Peterson Avenue
Chicago, IL 60659
Journal: *Journal of UFO Studies* (*JUFOS*)
Newsletter: *International UFO Reporter* (*IUR*)

The Center for UFO Studies was formed in 1973 by J. Allen Hynek, the head of Northwestern University's astronomy department, and Sherman J. Larsen, a businessman and director of a small UFO group in Chicago. During the 1950s and 1960s Hynek had been the U.S. Air Force's chief scientific consultant on UFOs, until he publicly complained that the military was doing a very poor job of investigating reports. While the prevailing air force attitude toward UFOs was one of disbelief and dismissal, Hynek thought that UFOs were probably more than mistaken identifications and hoaxes. CUFOS was created so that scientists and other trained professionals could deal with UFO research in a serious but open-minded way.

CUFOS is one of two major UFO groups in the United States (the other is the Mutual UFO Network; see MUFON entry below). CUFOS publishes a newsletter, the *International UFO Reporter* (*IUR*), and the *Journal of UFO Studies* (*JUFOS*), and has sponsored UFO investigations. Located in Chicago, its huge collection of research materials is available to people studying UFOs. The organization's official name was expanded after Hynek's death in 1986.

Mutual UFO Network, Inc. (MUFON)

103 Oldtowne Road
Seguin, TX 78155

Monthly magazine: *MUFON UFO Journal*

Walter H. Andrus, Jr., a former officer of Tucson's Aerial Phenomena Research Organization, founded the Midwest UFO Network in 1969, based in Quincy, Illinois. In 1975 his group moved to Seguin, Texas, and became MUFON, the Mutual UFO Network. One of the most successful UFO organizations in the brief history of UFOs, it would claim 4,000 national and international members by 1992. Though open-minded about different explanations for UFOs, MUFON clearly leans towards the extraterrestrial hypothesis. The organization hosts a conference in a different U.S. city every year. MUFON's monthly magazine, *MUFON UFO Journal* (formerly *Skylook*), contains serious UFO studies and is essential reading for ufologists.

Society for the Investigation of the Unexplained (SITU)

Box 265
Little Silver, NJ 07739

Magazine: *Pursuit*

Founded in 1965 by science writer and lecturer Ivan T. Sanderson, the Society for the Investigation of the Unexplained publishes the magazine *Pursuit* about three or four times a year. *Pursuit* reports on anomalies and looks for the explanations or meaning behind them.

Society for Scientific Exploration (SSE)

Box 3818, University Station
Charlottesville, VA 22903

Semiannual journal: *Journal of Scientific Exploration*

Semiannual newsletter: *Explorer*

The Society for Scientific Exploration was formed in 1982. It encourages scientific study of UFOs, unexpected animals, supernatural claims, and other subjects that lie on the edges of science, because "progress towards an agreed understanding of such topics ... is likely to be achieved only if they are subject to the normal processes of open publication, debate, and criticism which constitute the lifeblood of science and scholarship." Full SSE members must be connected with a major university, government group, or corporate research institution; those who do not qualify are associate members. The society sponsors a conference each year at an American university. It publishes both the newsletter *Explorer* and the *Journal of Scientific Exploration* twice a year.

INDEX

Bold numerals indicate volume numbers.

Folklore **3:** 410, 447-486
Foo fighters **1:** 4
Fort, Charles **1:** 62-63, 128-130, 132-133, 140-141, 168; **2:** 198, 234, 335, 340; **3:** 415, 486
Fortean Picture Library **3:** 482
Fortean Times **3:** 482
Fortin, Georges **1:** 151
The 4D Man **1:** 59
Fourth dimension **1:** 58-59, 94
Freedom of Information Act **1:** 14
Freeman, Paul **2:** 247-249
Friedman, Stanton T. **1:** 14, 74
Friedrich, Christof. *See* Zundel, Ernst
Frogs
    entombed **2:** 223
    falls from sky **1:** 130
From Outer Space to You **1:** 63

G

Gaddis, Vincent H. **1:** 81, 89, 94, 96, 106; **2:** 340
Galactic Federation **1:** 45
Garcia, Ignacio Cabria **1:** 52
Gardner, Marshall B. **1:** 55-56
GeBauer, Leo A. **1:** 75
Genzlinger, Anna Lykins **1:** 81
George Adamski Foundation **1:** 63
Ghost lights **1:** 103-110
Ghost rockets **1:** 4
Giant octopus **3:** 380-385
Giant panda **2:** 195, 287
Giants **2:** 325
Giant squid **2:** 195; **3:** 382, 388-393
Gimlin, Bob **2:** 245
Globsters **3:** 384-387
Gloucester, Massachusetts **3:** 396-397
Goldwater, Barry **1:** 73
Gorillas **2:** 226, 232, 234, 255
Gould, Rupert T. **3:** 414-416
Government conspiracies **3:** 495
Graham, Francis **1:** 64
Grain: falls from the sky **1:** 130
Gray, Hugh **3:** 419-420
The Great Flood **2:** 325; **3:** 484
The Great Sea-Serpent **2:** 195; **3:** 400
Green fireballs **1:** 111-114
Green, John **2:** 241, 244, 255-257
Green slime: falls from the sky **1:** 131
Greenwell, J. Richard **2:** 198-199, 209, 287-288, 308, 343-347; **3:** 387, 439
The Grey Selkie of Sule Skerrie **3:** 462
Griffiths, R. F. **1:** 139, 140

Group d'Etude des Phenomenes Aero-spatiaux Non-Identifies (GEPA) **1:** 11
Guiler, Eric R. **2:** 317-318

H

Hagenbeck, Carl **2:** 302
Hail **1:** 138-139
Hairy bipeds **2:** 197, 199, 231-292
Hairy dwarfs **1:** 41-42
Hall, Asaph **1:** 69
Halley's comet **1:** 164
Hall, Mark A. **3:** 478
Hallucinations **3:** 459
Hangar 18 **1:** 14, 73-80
Hangar 18 **1:** 75
Harry and the Hendersons **2:** 248
Hart, W. H. H. **1:** 18
Hazelnuts: falls from the sky **1:** 131
Herschel, John **1:** 60
Herschel, William **1:** 60-61
Hessdalen lights **1:** 106-110
Heuvelmans, Bernard **2:** 193, 195-197, 257-261, 272, 302, 304, 324; **3:** 393-394, 402-405, 469
Hillary, Edmund **2:** 268, 270
Himalayan Mountains **2:** 265
Hitler, Adolf **1:** 57
Hoaxes
    airships **1:** 19
    contactees **1:** 46
    crop circles **3:** 505
    hairy bipeds **2:** 240
    Loch Ness monster **3:** 421
    Minnesota iceman **2:** 257-261
    moon **1:** 61-62
    sea serpent **3:** 404
    space brothers **1:** 46-47
    spaceship crashes **1:** 75
    thylacine **2:** 320
    UFO **1:** 12-13
    Ummo **1:** 52
Holiday, F. W. **3:** 424, 520
Hollow earth **1:** 55-58
The Hollow Earth **1:** 57
Holmes, Sherlock **2:** 337
Homo sapiens **1:** 25
Hopkins, Budd **1:** 14, 167
The Hound of the Baskervilles **2:** 337
Hudson, Henry **3:** 467
Human-animal transformations **3:** 470-471
Humanoids **1:** 11, 15, 37-40, 41-42, 153
Hynek, J. Allen **1:** 7-9, 50, 107, 173, 178-179; **3:** 458

Q

Queensland tiger **2:** 213-214

R

Randle, Kevin **1:** 74
Red wolf **3:** 471
Regusters, Herman **2:** 296, 308-309
Religions **1:** 25
Religious lights **1:** 105-106
The Report on Unidentified Flying Objects
  (1956) **1:** 6
Reptile men **2:** 199, 329-331
Return of the Ape Man **2:** 239
Ri **2:** 198, 345-347
Ridpath, Ian **1:** 32-33
Road in the Sky **1:** 30
The Robertson Panel **1:** 6-7
Robins Air Force Base **1:** 5
Rocky Mountain Conference on UFO
  Investigation **1:** 47
Rogo, D. Scott **1:** 125
Rojcewicz, Peter M. **1:** 50
Roswell incident **1:** 14, 73-80, 171
The Roswell Incident **1:** 66
Rudkin, Ethel H. **2:** 338
Runaway clouds **1:** 150
Ruppelt, Edward J. **1:** 5, 6, 113

S

Sagan, Carl **1:** 32, 33
Salamanders: falls from the sky **1:** 131
Sananda **1:** 13
Sanderson, Ivan T. **1:** 81, 96, 98-99; **2:**
  196-197, 233-234, 244, 255, 257-261,
  267, 297-298, 304; **3:** 379
Sandoz, Mari **3:** 448
Sarbacher, Robert **1:** 78-79
Sasquatch **2:** 241-242
Satan worshippers **3:** 494-495
Saucer clouds **1:** 154
Saucer nests **3:** 503-504, 506
Saucers, flying **1:** 3-16
Sauropods **2:** 295, 302, 310, 323
Scandinavia **3:** 413
Schmidt, Franz Herrmann **2:** 298-300
Schmitt, Don **1:** 75
Science Frontiers **1:** 141
*Scientific Study of Unidentified Flying
  Objects* **1:** 9
Scully, Frank **1:** 75

Sea cows **3:** 426, 468
Sea monsters **3:** 375-380
Searching for Hidden Animals **2:** 307
Sea serpents **2:** 195; **3:** 394
Secret of the Ages **1:** 55, 57
Sedapa. *See* Orang-Pendek
Seeds: falls from the sky **1:** 131
Selkies **3:** 462
Serpents **2:** 322, 323; : 394-405; 447-452
Sesma, Fernando **1:** 50, 52
Seventh District Air Force Office of
  Special Investigations (AFOSI) **1:** 111
Shape-changing **2:** 337; **3:** 435, 473
Shaver, Richard Sharpe **1:** 57
Shiels, Anthony "Doc" **3:** 404, 421
Shooting stars **1:** 112
Siegmeister, Walter. *See* Bernard,
  Raymond
Sirius **1:** 31
Sirius B **1:** 31
Sirius mystery **1:** 31-33
The Sirius Mystery **1:** 31-33
Sirrush **2:** 322-324
"Sky gods" **1:** 30
Skyquakes **1:** 160-161
Sky serpents **3:** 447-452
Slick, Tom **2:** 268-270, 273
Smith, J. L. B. **3:** 449
Smith, John **3:** 466
Smithsonian Institution **2:** 193, 258-259
Smith, Wilbert B. **1:** 78
Society of Space Visitors **1:** 50
Solar system **1:** 68
Somebody Else Is on the Moon **1:** 64
Sonar **3:** 413, 425, 432
South Pole **1:** 57
Space animals **1:** 133; **2:** 340
Space brothers **1:** 12, 45-47
Space Review **1:** 48
Spencer, John Wallace **1:** 96
Sperm whales **3:** 391, 393
Spielberg, Steven **1:** 9, 79, 92
Splash **3:** 465
Spontaneous generation theory **1:** 121
Spontaneous human combustion
  3: 489-493
Sprinkle, R. Leo **1:** 47
Squids **3:** 388-393
Squid-whale battles **3:** 391-393
Star jelly **1:** 142-145
Starman **1:** 47
Steckling, Fred **1:** 63
St. Elmo's fire **1:** 156
Stewart, Jimmy **2:** 270

Stonehenge **3:** 499
Stones: falls from the sky **1:** 123
Strange cats **2:** 205-214
Strange clouds **1:** 149-155
Strange Harvest **3:** 496
Strange Phenomena **1:** 141
Strange rain **1:** 130, 134-136
Stranger Than Science **1:** 96
Strategic Defense Initiative **1:** 84
Straw: falls from the sky **1:** 132
Strieber, Whitley **1:** 167
Stringfield, Leonard H. **1:** 14, 75, 77
Submarines **1:** 98; **3:** 378
Sulphur: falls from the sky **1:** 123
Sulphur Queen **1:** 94-95
Sumatra **2:** 278
Summers, Ian **3:** 497-498, 517
Sunspots **1:** 69
Supernatural phenomenon **1:** 103;
   3: 458
Super-Sargasso Sea **1:** 132
Surrcy puma **2:** 210-211
Symmes, John Cleves **1:** 55

T

Tasmanian devil **2:** 317
Tasmanian Tiger **2:** 315
Taylor, Charles **1:** 89-94
Telepathy **1:** 52
Teleportation **1:** 133; **2:** 212
Temple, Robert K. G. **1:** 31-33, 168
Theosophy **1:** 26
*They Knew Too Much about Flying
   Saucers* **1:** 49
Thunderbird photograph **3:** 481
Thunderbirds **3:** 476-483
Thunderstones: falls from the sky **1:** 126
Thylacine **2:** 198, 315-322
Thylacoleo **2:** 214
Tibet **2:** 265
Tien Shan, China **2:** 281
Titmus, Bob **2:** 249
Toads: entombed **2:** 223-226
Toads: falls from the sky **1:** 128
*Tornados, Dark Days, Anomalous
   Precipitation, and Related Weather
   Phenomena* **1:** 149
Totem poles **3:** 476
Transient lunar phenomena (TLP)
   1: 59-66
Traverspine gorilla **2:** 232
Trench, Brinsley le Poer **1:** 55, 57
Tunguska event **1:** 162-165

Turtles: falls from the sky **1:** 132
20,000 Leagues Under the Sea **2:** 197;
   3: 378, 391

U

The UFO Experience **1:** 10
Ufology **1:** 16
UFOs **1:** 3-16
   abductions **1:** 15, 16
   alien corpses **1:** 75
   and black dogs **2:** 339
   and cattle mutilations **3:** 496
   and clouds **1:** 152-153
   and crop circles **3:** 503
   and fairy sightings **3:** 458, 459
   and flying humanoids **1:** 40
   and the fourth dimension **1:** 59
   and hairy bipeds **2:** 237
   and hairy dwarfs **1:** 41-42
   hoaxes **1:** 12, 16
   and Mothman **2:** 335-336
   and ocean civilizations **1:** 98-99
   photographs **1:** 12
   and USOs **3:** 378
   wreckage **1:** 73
UFOs: Operation Trojan Horse **1:** 59;
   2: 335
Ultraterrestrials **1:** 59; **2:** 335
Ummites **1:** 52
Ummo **1:** 50-52
Underground civilizations **1:** 55-58
Underwater civilizations **1:** 96, 98-99 **2:**
   233; **3:** 379-380
Unidentified airships **1:** 17-24
Unidentified flying objects. *See* UFOs
Unidentified submarine objects (USOs)
   3: 375-380
Uninvited Visitors **2:** 233
Unknown Animals **2:** 198
Unveiled Mysteries **1:** 56
Uranus **1:** 66
U.S. Air Force **1:** 4-9
U.S. lunar explorations **1:** 64

V

Vallee, Jacques F. **1:** 51; **3:** 458
Vandenberg, Hoyt S. **1:** 5
Vanishing clouds **1:** 154
Venus: visitors from **1:** 63
Verne, Jules **2:** 197; **3:** 378

Vile vortices **1:** 98-99
Vogel, W. J. (Bill) **1:** 107
Von Däniken, Erich **1:** 25-26, 30, 168
Vulcan **1:** 66-69

W

Waddell, L. A. **2:** 265
Wagner, Roy **2:** 345
Wandering kangaroos **2:** 203-205
Water horses **3:** 410-411, 414
Waterspouts **1:** 121-122
Webb, DeWitt **3:** 380, 382-383
Webster, Daniel **3:** 398
We Discovered Alien Bases on the Moon
    1: 63
Werewolf of London **3:** 474
Werewolves **2:** 199, 217-220
White Pongo **2:** 232
White River monster **3:** 442
Wild men **2:** 198, 281, 285
Wild Talents **2:** 234
Williamson, George Hunt **1:** 26, 30
Wilson, Don **1:** 64
Wilson, Robert Kenneth **3:** 419, 421
Winer, Richard **1:** 96
Witchcraft **1:** 95
Witiko **2:** 242

Wolfman **2:** 219
The Wolf Man **3:** 474
Wolves **2:** 217-220; **3:** 470-474
World War II **1:** 4, 57, 80
World Wildlife Fund **2:** 317
Wright, Bruce S. **2:** 207
Wright Field **1:** 8
Wright-Patterson Air Force Base **1:** 4, 6,
    75-80, 79

Y

Yahoo **2:** 289-291
Yakima **1:** 107, 108
Yakima Indian Reservation lights
    1: 106
Yeren **2:** 284-289
Yeti **2:** 198, 244, 265, 273, 282, 284
Yowie **2:** 289-292

Z

Zarzinski, Joseph W. **3:** 427, 431-433
Zeuglodons **3:** 412, 426, 438-439, 441
Zhamtsarano, Tsyben **2:** 282
Zhou Guoxing **2:** 286-287
Zundel, Ernst **1:** 57